Internet of Things: Concepts and System Design

Milan Milenkovic

Internet of Things: Concepts and System Design

 Springer

Milan Milenkovic
IoTsense
Dublin, CA, USA

ISBN 978-3-030-41348-4   ISBN 978-3-030-41346-0   (eBook)
https://doi.org/10.1007/978-3-030-41346-0

This Springer imprint is published by the registered company Springer Nature Switzerland AG
The registered company address is: Gewerbestrasse 11, 6330 Cham, Switzerland

Preface

Internet of Things (IoT) systems deploy smart connected things that sense their environment and generate quantitative data about the physical world. In effect, IoT adds a new dimension to the Internet, awareness of the real world. This transformational change bridges the gap between physical and virtual/cyber worlds that has persisted since the invention of computing. Implementations of IoT technology are expected to create tremendous opportunities for new uses and applications. Their business impact is projected to be in the trillions of US dollars, comparable to the size of world's major economies.

The design of IoT systems spans multiple disciplines and requires diverse skills and knowledge in a number of areas. They include sensors, embedded systems, real-time systems, control systems, communications, protocols, Internet, cloud computing, large-scale distributed processing and storage systems, AI, and ML, coupled with the domain experience in the areas where they are to be applied, such as building management or manufacturing automation. Obviously, it is not reasonable to attempt to cover all those disciplines in detail in any single text.

This book is devoted to the design principles and practices for implementing IoT systems. Its primary purpose is to provide a foundation and a reference for students and practitioners to build upon when analyzing and designing IoT systems, as well as to understand how the specific parts they are working on fit into and interact with the rest of the system.

This book provides a comprehensive overview of the IoT systems architecture, including an in-depth treatment of all key components. The emphasis is on a complete and balanced treatment at roughly equal level of depth for all covered topics. The exposition takes a system's approach by describing the big picture and how the overall system works so that readers can interpret subsequent in-depth topical coverage in context. In addition to describing the functional and foundational aspects of all major IoT system components, the emphasis is on integration, i.e., how the components and building blocks are combined to create complete IoT systems. The book is organized as follows.

Chapter 1 describes the transformational impact and importance of adding the IoT real-world sensing dimension to the Internet. It also outlines key differences between IoT systems and the Internet that pose some unique design requirements and challenges. Chapters 2, 3, 4, 5, and 6 treat the foundational aspects of IoT systems and design.

Chapter 2 covers the edge, starting with sensor data acquisition and processing, and continues with the edge functionality that includes event processing, storage, local control and scripting, and interfacing to sensors and actuators as well as to the external communications and the cloud. It discusses the tradeoffs involved in the functional placement of components in distributed systems on the edge-to-cloud continuum, including the fog, and concludes with a description of hardware and software considerations involved in the edge node design.

Chapter 3 focuses on communications. It describes a layered network design which is the underpinning of the Internet and a useful blueprint for the IoT system design. It continues with the coverage of wireless and constrained networks at the edge, followed by the IoT adaptations of the cellular networks in the licensed spectrum. The chapter concludes with the exposition of constrained protocols and the messaging and queuing publish-subscribe mechanisms designed for IoT systems.

Chapter 4 focuses on the cloud. It covers key elements and functions of the IoT cloud core components, including data ingestion via edge-cloud gateways, in-flight stream processing, and short- and long-term storage systems suitable for IoT applications. The rest of the chapter covers analytics and principles of machine learning, operation of artificial neural networks, and types and uses of ML systems.

Chapter 5 covers the control plane, security and management systems. It describes the types of security threats and attacks in IoT and OT systems, followed by the security analysis and planning steps. A section on cryptography overviews key foundational elements of security design, including symmetric and public-key cryptography. This is followed by the treatment of endpoint security, including hardware security and software isolation mechanisms, network security, and privacy.

Chapter 6 is devoted to the topic of data representation and semantic interoperability, which is a key new design requirement that IoT systems introduce. It describes approaches to addressing the issue, which is a prerequisite for enabling big IoT data aggregations for meaningful insights and effective applications of ML and AI techniques.

Chapter 7 contains an overview of representative IoT standards dealing with data and information models. Chapter 8 outlines key components and design choices of several commercial IoT platforms. These chapters illustrate how the underlying principles of IoT system design may be reduced to practical instantiations that can serve as the potential building blocks in IoT system designs.

Given the book's focus on system design and integration, most chapters contain the "putting it all together" sections to indicate how the concepts may be put together. Chapter 9 is devoted to system-level integration in its entirety. It also presents a detailed example of a complete IoT system, including conceptual design, implementation, results, post deployment user studies, and a discussion of tradeoffs and issues encountered in the process.

Dublin, CA, USA   Milan Milenkovic

Acknowledgments

Interactions with a number of people have clarified and shaped many of the ideas and concepts described in this book. They include, in alphabetical order, Tomm Aldridge (Intel), Ajay Bhatt (Intel), Ken Birman (Cornell University), Carsten Bormann (Universität Bremen), Simon Crosby (swim.ai), David Culler (University of California Berkeley), Stephen Dawson-Haggerty (Comfy), Brian Frank (SkyFoundry), Jaime Jimenez (Ericsson), Jim Kardach (Intel), Randy Katz (University of California Berkeley), Ari Keränen (Ericsson), Michel Kohanim (Universal Devices), Michael Koster (Samsung SmartThings), Matthias Kovatsch (Huawei), Andrew Krioukov (Comfy), Rick Lisa (Intel), Brian McCarson (Intel), Michael McCool (Intel), Florian Michahelles (Siemens), Doug Migliori (ControlBeam), Alexandros Milenkovic (Curia), Neal Mohammed (Rudin Management), Jorge Ortiz (Rutgers University), Dirk Pesch (University College Cork), John Petze (SkyFoundry), David Prendergast (Maynooth University), Kumar Ranganathan (Intel), Eve Schooler (Intel), and Sven Schrecker (LHP Engineering Solutions).

Major contributors to the POEM project described in Chap. 9 include (from Intel) Yves Aillerie, Ulf Hanebutte, Catherine Huang, Sailaja Parthasarathy, Han Pham, Sylvain Sauty, Scott Shull, and Jun Takei. Executive support from Lorie Wigle (Intel), Marie Annick Le Bars (Bouygues Immobilier), and Renaud Deschamps (Lexmark) made it possible.

All errors and omissions are mine.

Contents

Chapter 1
Introduction and Overview

Internet of Things (IoT) systems connect the physical world to the Internet. Basically, IoT works by attaching real-world interfaces to the Internet, such as sensors that provide data and actuators that act upon their surroundings. In effect, IoT systems provide the technology and means to instrument, quantify, and actuate the physical world.

Connecting sensors adds physical-world data, and in a sense awareness, to the Internet. This addition is a transformational change; it basically bridges the gap between physical and virtual/cyber worlds that has persisted since the invention of computing. In effect, IoT augments the Internet with all its features and capabilities by adding to it the real-world dimension. With the incorporation of IoT, the Internet becomes a web of people, information, services, and things, essentially the Internet of everything. As Kevin Ashton, who coined the term Internet of Things points out "In the twentieth century, computers were brains without senses – they only knew what we told them. In the twenty-first century, because of the Internet of Things, computers can sense things for themselves" [1].

There is no formal and commonly accepted definition of IoT. Many attempts tend to be descriptive and list system attributes, such as "a global infrastructure for the information society, enabling advanced services by interconnecting (physical and virtual) things based on existing and evolving interoperable information and communication technologies" [2] or "the extension of Internet connectivity into physical devices and everyday objects. Embedded with electronics, Internet connectivity, and other forms of hardware (such as sensors), these devices can communicate and interact with others over the Internet, and they can be remotely monitored and controlled. The IoT adds the ability for IoT devices to interoperate with the existing Internet infrastructure" [3].

Analysts predict that the installed base of connected devices will number in tens of billions in the next few years, and the number of sensors is projected to grow to hundreds of billions in the near future [4]. In the same time frame, the business

© Springer Nature Switzerland AG 2020
M. Milenkovic, *Internet of Things: Concepts and System Design*,
https://doi.org/10.1007/978-3-030-41346-0_1

impact of IoT is projected to be in trillions of US dollars, comparable to the GDP of the world's largest economies. Regardless of how projections turn out, those are staggering numbers indicating a truly transformational change.

Connection of real and cyber worlds using the Internet fabric and protocols enables many new types of hybrid interactions and thus the potential for the creation of the plethora of exciting new uses, applications, and business models. Its development holds the promise to profoundly impact not only the industry but also many aspects of the daily lives and well-being of people.

IoT Systems

Figure 1.1 illustrates the major functional components of an IoT system. At the bottom are pictorial representations of some of the applications of IoT, including office, automotive, transportation, manufacturing, agriculture, home, office building, and smart cities. The layer above illustrates some representative categories of IoT sensors and things.

Some sensors and things are designed to connect directly to the Internet and communicate with applications and services residing in the cloud. Examples include a variety of devices, often marketed with a moniker smart, such as security cameras, fire sensors, thermostats, appliances, and power meters.

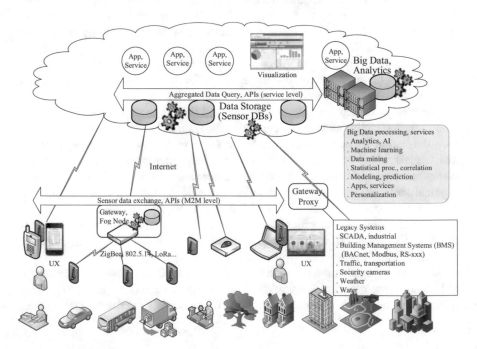

Fig. 1.1 IoT systems

Other sensors connect to the rest of the IoT system and the Internet using intermediaries, such as gateways. Gateways and fog nodes are usually more powerful devices connected to collections of usually low-end sensors attached to it via local network links, often wireless, such as ZigBee, variants of 802.5.14 networks, Bluetooth, and low-power Wi-Fi. Gateways usually provide wide-area connectivity and edge processing for the attached sensors that may come in the form of protocol conversion, data storage and filtering, event processing, and analytics.

Billions of smart phones already in user's hands play multiple useful roles in IoT systems. They can act as smart connected sensors, gateways, and user interaction devices. Phones have a variety of built-in sensors – such as fingerprint, pressure, light, Hall sensor, barometer, temperature, geomagnetic, accelerometer, gyro, proximity, and global positioning system (GPS) – that can be used to report device location and state. These sensors are program-accessible by applications and therefore connected or connectable to the Internet. Many other location-based and augmented reality (AR) applications are possible by combining these with data from other phone sensors, such as cameras and microphones.

Smart phones and tablets may also act as a form of IoT gateways for other devices on their reachable local networks, such as multitudes of pairable Wi-Fi- and Bluetooth-enabled sensors including fitness trackers, smart watches, cars, and home automation devices. In Fig. 1.1, smart phone types of devices are shown as being both connected sensors and user interaction devices. Smart phones can accept user input and visualize outputs from applications, provide user interfaces for data and control inputs, as well as visualize any IoT data of interest available locally or anywhere else on the Internet.

IoT endpoints and sensors may also be embedded into the existing IT infrastructure and devices, such as personal computers. Adding IoT sensors and interfaces to existing IT devices, such as PCs and laptops, can provide coverage close to users and lower system cost by capitalizing on the host's power, connectivity, processing, and storage resources. In addition, a PC can act as a gateway and a rich user experience (UX) endpoint in an IoT system.

In addition to sensors and devices designed expressly for IoT uses, there are numerous sensors collecting real-world data in various forms of legacy systems such as industrial automation, energy and health systems, building management systems (BMS), and supervisory and control systems (SCADA). For example, a BMS may contain thousands of physical-world sensors, such as temperature and motion in each room of an office building. Most of them are locked in proprietary formats, but BMS and SCADA systems are increasingly becoming interfaced to the Internet at their top levels of hierarchy for at least restricted forms of sharing and remote control. Data from such systems can be incorporated into IoT systems for aggregation and processing purposes with the aid of protocol and data format translators, often implemented as purpose-built gateways.

"Back-end" processing, often performed by remote servers in IoT systems, is depicted by a generic cloud in Fig. 1.1. The picture is not intended to imply that there is a single centralized cloud in the IoT space, there are actually many, but rather to highlight some of the key components of an IoT system that support processing of data from large numbers of sensors by a variety of applications. Conceptually, the cloud is usually the top layer in an IoT system hierarchy, where data from a variety of diverse sources are aggregated and processed for optimization and discovery of global trends and relations. Depending on their nature and real-time requirements, incoming sensor data and events may be processed "in-flight" as streams, stored for post-processing and archival purposes, or both. In addition to supporting and running applications, an IoT cloud may also contain some common services such as large-scale storage, analytics-processing engines, data visualization and graphing, as well as management functions such as security and provisioning, not shown in Fig. 1.1.

Machine learning (ML) and artificial intelligence (AI) algorithms are usually operated in the cloud where they can work with large aggregations of data. In current practice, such algorithms tend to perform better on large data sets that improve training and consequently their prediction capability. The value of IoT data aggregations increases significantly if they can acquire and store data in an interoperable format, thus supporting better analytics on larger data sets amassed from multiple sensing domains.

In terms of typical IoT data flows, sensors monitor physical-world status and provide the corresponding digital readout of those, sampled at some data rate that may be fixed or variable through command settings. Flows of bits in IoT systems can be bidirectional, with sensor data typically flowing upstream to the cloud and towards applications and control and status setting flowing downstream, such as a command to open a valve.

In addition to data capture, sensor streams need to be tagged with metadata to provide additional information about the meaning of data and the context in which they were acquired. Metadata may indicate the nature of the data, e.g., temperature, reporting location, ownership, and structural relations. Metadata may be added to sensor data streams anywhere on the path from the point of capture to sensor data storage in the cloud and in different stages of the lifecycle, including installation, provisioning, operation, and management.

Although not explicitly indicated in Fig. 1.1, applications that process and react to sensor data may reside at the edge, in the cloud, or distribute their functionality across multiple system components. For instance, implementation of a fitness application can be a fairly complex IoT system. It starts with a wearable device that may include a variety of sensors to detect a user's activity, motion, position, and perhaps some vital indicators such as heart rate or EKG. This can be a complex embedded system that needs to constantly acquire and preprocess data, have a long battery life, and be as small as possible. It usually contains some preliminary processing functionality, such as data capture, preprocessing, time stamping, and buffering of data until its transmission to the phone or cloud is completed. Additional computation and storage may be placed in the portion of the application running on the smart phone that the sensor is paired with. In addition, the phone application can provide a user interface to visualize the data and customize settings such as individual targets. The phone also typically acts as a gateway by relaying user data to the cloud portion of the fitness application and service that can archive user data and provide more comprehensive types of analyses, such as trending and comparisons of results with a user's friends groups. In addition, the cloud application can aggregate hopefully anonymized data from its entire user base that may be used to study population trends for various uses, such as user reference and medical studies.

Figure 1.1 illustrates data and control flows related to the primary functional aspects of operation of IoT, namely, collection of data and processing to act on it. This is sometimes referred to as the data plane or user plane. Another important dimension of an IoT system, not shown in Fig. 1.1, is the control plane that maintains security and operational aspects of the IoT infrastructure itself. Its primary components are security and device management that include addition and bringing up of new devices and operating the system safely and reliably through constant monitoring and remediation when necessary.

Why Now?

A confluence of technological and infrastructure developments centered around the Internet forms much of the basis and impetus for the construction of IoT systems. The major ones include:

- Industry 4.0 and digitalization
- Sensors – installed base, variety, lower cost, and easier integration
- Smart phones – sensors, gateways, and UI devices
- Cloud computing – global and capacity on demand
- AI and ML technologies – actionable insights with IoT data
- Internet – technology, global infrastructure, and users

Industry 4.0 generally refers to the digital transformation of manufacturing using a combination of technologies including automation, robotics, digital-to-physical transfers such as 3D printing, big data, ML and AI, robotics, and augmented reality (AR). Some of its key foundations are machine-to-machine communications, Internet compatibility, and interoperability. With sensing and digitization of data as its starting point, Industry 4.0 shares many ingredients with Internet of Things, and IoT is generally regarded as one of its foundational technologies. Enterprise-scale investments in Industry 4.0 will facilitate deployment and accelerate the industrial dimension of the IoT technology evolution. One of the aspects of Industry 4.0 is bridging and format translation to incorporate the data already being collected by the large installed base of sensors in legacy systems mentioned in the previous section.

The availability of a variety of connectable *sensors* in the consumer space facilitates construction of IoT systems and enables many potential applications. Home automation and monitoring devices, such as smart thermostats, lights, security systems, and Internet-enabled cameras, can perform useful functions in isolation and be combined to perform more complex functions by engaging in coordinated ensemble behaviors. A growing base of personal sensors, such as wearables, enables the continuous tracking of physical activities and vital signs for fitness and health applications.

Another major contributor of inexpensive and widely available sensors that are comparatively quick and easy to deploy is the rapidly growing "maker" movement [5, 6]. It has brought about inexpensive processor boards for prototyping IoT systems, such as Raspberry Pi [7], and commonly used and de facto standardized I/O interfaces, such as GPIO (general-purpose input/output) and packaging in form factors such as Arduino shields [8]. This has led to the availability of a wide range of compatible sensors and actuators – such as temperature, light, humidity, pressure, presence, air quality, touch, distance, current, vibration, dust, heart rate, sound, switches, relays, and motor drivers – whose cost tends to be orders of magnitude lower than their traditional industrial counterparts [9]. These sensors usually come • with drivers and libraries for popular development environments, such as Arduino. Prior to these developments, constructing of a hardware prototype to test an idea with potential users required costly investment into custom development, capital expenditures, and delays in scheduling the generally less profitable small production runs of product prototypes. The availability of low-cost sensors, platforms, and software tools drastically reduces the time, costs, and risks of development, testing, and experimentation with IoT prototypes. This spurs innovation by allowing rapid and inexpensive cycling and testing of variants to validate a product and its features before committing resources to its production.

As described in the previous section, *smart phones* provide a large installed base of smart and connected sensors that can be used in a variety of applications. Phones can also act as gateways between sensors in the user's proximity and applications and services residing locally or anywhere in the global cloud. In addition, phones provide user interfaces for data and control inputs, as well as visualization of virtually any IoT data and relevant content on the Internet.

The availability of *cloud computing* facilitates the construction and deployment of an IoT system with global reach and without the requirement for up-front capital investment in servers, cloud networking, and storage. Moreover, it comes with the appearance of almost instant and limitless scaling of capacity to meet variations in demand. Cloud technology and commercial offerings provide the capability to meet IoT system requirements in terms of volume and bandwidth for large-scale data ingestion, real-time event processing and Internet-scale databases, and storage for large aggregations of data.

Advances in *AI and ML technologies* have led to major successes in applications such as image processing and natural language understanding and translation. There are growing expectations that similar achievements may be accomplished by applying AI and ML techniques in IoT systems to obtain insights leading to actionable outcomes. There are many AI and ML tools, some of them with libraries of neural network configurations with pre-trained parameters for specific applications that may be used as a starting point to accelerate IoT experimentation and adaptation. Many of these tools are available on the Internet and hosted by cloud providers.

The most important component of IoT systems is their foundation and key pillar, the *Internet* itself. We use the term Internet in a broad sense that includes not only its original underpinning of IP-enabled networks and protocols but also the World Wide Web and numerous applications that it has enabled. Internet size, vibrancy, global reach, services such as public clouds, users, development tools, and an army of developers provide a formidable foundation for IoT systems to add to and build upon.

The Internet provides a set of standardized communication protocols and data markup languages that allow the exchange of data by endpoints and their use by multiple applications. Standardization and adherence to common specifications enable independent design of components that are modular and interoperate. As a result, a user can select a device of their choice from one vendor, pick a browser from another source, and use the combination to access Internet-compliant content posted anywhere in the world. The content itself may be created using any compliant authoring tool supplied by a variety of vendors and hosted on any of a variety of web servers that may run on any of a number of different hardware and software platforms.

Internet technology has been proven in the construction and operation of immensely scalable, (almost) always available systems with the ability to handle millions of servers and billions of users while providing global reachability to both producers and consumers of data. Moreover, the huge size of its already installed base and global infrastructure that is already deployed provides a nearly ubiquitous availability of the Internet in many of the populated parts of the world. Its modularity and adherence to standards have spurred innovation and created a competitive marketplace with many hardware and software offerings that can be mixed and matched to create services and solutions.

How IoT Systems Are Different?

While being able to capitalize on much of the technology and infrastructure of the Internet, IoT systems have some characteristics that are unique and structurally different and require specialized design and engineering to implement them. Much of the rest of this book is devoted to those topics.

Some of the key IoT differences and challenges include:

- Information models, standards, and data and metadata formats and semantics
- Search
- Topology, inverse client to server ratios, and reverse data flows
- Continuous, time-series data streams
- Distributed function placement – edge, fog, and cloud

In order to illustrate some of the differences, consider a simplification of the common usage mode of the Internet to access a web page from one of the billions available on the web. The (human) user may start by searching for a phrase of interest or a partial name of the source. The search engine of choice typically returns a number of clickable links (URLs) to pages that contain content that is likely relevant. The page chosen by the user is displayed with the appearance and format created by its authors. It can also contain active links to other relevant pages that the user can read and navigate.

An important objective for IoT systems is to enable something of comparative functionality and power for IoT data. In order to provide a similar degree of flexibility that we came to expect on the Internet, IoT systems need to grapple with the difficult problem of machine-level understanding, i.e., interpretation of data generated by sensors and things.

Note that in the previous description of the Internet, data are destined for consumption and interpretation by humans. The Internet provides a method and protocols to transport data from servers to clients and use some annotation, such as HTML, to annotate the data for proper rendering at the receiving end. Rendering implies displaying the text with the author-specified appearance, such as the font type, and proper placement of images and other types of media. But beyond essentially mechanical rendering, there is no semantic interpretation of the data by the receiving client. The meaning of the data is defined by the natural language in which the text is authored. If you do not understand that language, no semantics is conveyed. As an example, look up a web site of a major newspaper in a language that you cannot read or do not speak (in my case, Japanese or Arabic). More precisely, the textual Internet provides the syntactic and the structural interoperability between the endpoints, but not the semantic one.

In IoT systems, data exchanged between machines at endpoints are sensor readings and actuator commands. These are typically device- and domain-specific numbers such as temperature readings, with annotations such as the nature of measurements and units of measure, and control strings that do things like change a set point. Applications (not humans) at the receiving end need to be able to "understand" and interpret messages in the payload in order to process them appropriately. For this to work, there has to be some common understanding between the senders and the receivers on what the data means and how to encode and annotate it for transmission and to correspondingly decode it upon reception.

The problem here is that there is *no commonly accepted convention for representing data in IoT systems*. In its absence, it is not possible to independently develop IoT endpoints that would interoperate as is common on the rest of the Internet. A number of standardization efforts are under way to address this problem. The challenge faced by IoT standards is to replicate the Internet benefits of modularity and interoperability to data and control exchanges at the edge. This means that components may be developed independently, on different platforms and by different establishments, and reasonably expected to be able to connect and interoperate with others provided they correctly implement the applicable standards. IoT data representation is an important design problem, and we devote two separate chapters of this book to IoT data definition, interoperability, and work in standards that are addressing it.

Internet *searching* techniques are not directly applicable to IoT since its "content" is not the textual information in a natural language that can be indexed by the common web crawlers. Consequently, it is not possible to search the IoT space for content based on names or attributes of the endpoints. Moreover, many of IoT endpoints do not have Internet resolvable names for reasons of scale, domain-name assignments and, in some cases, are kept intentionally private for security. Such systems may deploy mechanisms of device self-description and discoverability or maintain dedicated device directories to enable limited, domain-specific searches. In any case, in order to enable attribute-based searching, IoT data need to be annotated by contextual and semantic information, usually in the form of metadata.

Traditional Internet servers tend to be located closer to the core of the network for higher bandwidth and to reduce access latencies. Clients with browsers tend to be more towards the edge of the network. The current statistics suggest that the number of servers on the Internet is on the order of 100 million and the number of clients is closer to 4 billion. The bulk of data flows tends to be from the cloud core towards the edge as servers deliver requested data and media streams. This is partially evidenced by the fact that many edge connections, such as residential Internet links, tend to have asymmetric speeds, with downstream bandwidth often exceeding the upstream by an order of magnitude.

IoT systems are also different in terms of *topology, client to server rations, and primary direction of data flows.*

IoT nodes that source IoT data are referred to as servers. In practice these are usually the edge nodes and things that may have constrained capacity, power, connectivity, and bandwidth. Clients are applications and services that consume IoT data. They may reside on peer edge nodes, in gateways, in fog nodes, or in the cloud. Thus, unlike on the Internet, clients in IoT systems may have higher processing and storage capacity than IoT servers.

In an IoT system, the total number of servers with data sources, i.e., billions of endpoints, largely outnumbers the clients, i.e., IoT services and applications that consume them. An IoT server will typically provide data to a few clients or publish them to a messaging broker where any number of subscribed clients can use them. IoT clients implemented as cloud services or data aggregators may consume data from numerous IoT servers. The bulk of IoT data flows is from the edge servers towards the processing clients that are located closer to the system core.

Thus, in comparison with the mainstream Internet, IoT systems tend to have inverted client to server ratios and reverse data flows, from the edge towards the more centralized parts of the system where applications and services (clients) that operate on combinations and aggregations of data tend to reside.

Much of IoT data have *time-series* characteristics in the sense of being consecutive, time-stamped samples of sensor readings. Applications tend to retrieve such data by combinations of stream identities and periods of time, such as within the last hour or at a specified date. Volumes of IoT data can be very high as they are generated by machines and often continuously sampled, sometimes with very high frequency. IoT data may also have real-time processing requirements that may impose tighter bounds on latency tolerances. Thus, IoT system architecture and implementations need to account for potentially high volume and throughput of data ingress from the edge. In addition, they need to meet processing and storage requirements of time-series and real-time data.

Placement of functions is the quintessential design decision in distributed systems, such as IoT, where data and processing in the form of services and applications may not be co-located. Basically, it is a trade-off that revolves around bringing the data to processing or processing to data. Considerations driving the decision include processing capacity, storage, latency, connectivity, bandwidth availability, and cost. In IoT systems additional considerations may include real-time and local autonomy requirements in the sense that some degree of control functions should remain operational even when cloud access is not available.

In IoT applications where data volumes are high but significance of much of it is low, it is often better to process the raw data close to the source. For example, recognizing a human shape in a surveillance application may require constant acquisition of a video stream with potentially high bandwidth. However, since only fragments of interest are the ones with human shapes in them, considerable reductions in bandwidth may be realized by processing the video stream at the edge and forwarding data to the cloud only when the local AI infers a significant event, i.e., detection of a shape. In this example, potential savings in bandwidth have to be weighed against the cost and complexity of placing and managing processing capacity at the edge.

In principle, in IoT systems processing and storage functions may be placed anywhere along the data path, from the acquisition at the edge and traversal of intermediate processing and staging nodes, such gateways and fog, all the way to the cloud. Moreover, processing modules may be distributed at various points on the data path to form the processing pipelines to complete increasingly more complex functions. Judicious placement of processing and storage functions is one of the important factors in IoT system design.

Value and Uses of IoT

What are the benefits and value that can be expected to be realized by deploying and using IoT systems? A somewhat oversimplified answer is provided by the old management adage – what you can measure, you can improve. IoT enables continuous sensing and measurement of the state and behavior of the physical world in order to react and ultimately improve on some of its aspects that can be controlled. Pervasive and continuous monitoring provides insights and a quantified view of the physical world. The extent and type of the resulting influence vary considerably depending on the nature of the system under observation. A manufacturing operation may be directly controlled by an IoT system using guidance provided by an AI system. On the other hand, continuous monitoring of a person's EKG via a smart personal sensor can identify and record anomalies and changes in vital signs that preceded it to be made available to a physician for diagnosis informed in a manner that was previously not possible.

Two primary areas of potential application of IoT systems are in business and consumer segments. Some of the early IoT consumer applications are already becoming familiar – such as self-driving cars, wearable health and fitness sensors, and Internet-enabled home security cameras and smart thermostats. However, the major financial impact is initially expected in business uses. In business processes, use of IoT can lead to transformational changes that impact efficiency and profitability of enterprises, such as improvements in predictive maintenance, better asset utilization, and higher productivity.

In the remainder of this section, we describe several applications to illustrate some of IoT uses that are under way or being contemplated. Many other current and potential uses of IoT systems are described, some in considerable detail, in other chapters in this book.

According to analysts, the ranking of IoT uses based on projected business value and financial impact may look as follows:

- Production environments – factories
- Cities
- Human health and fitness
- Retail environments
- Transportation and automotive
- Oil and gas and mining
- Utilities and energy
- Home
- Offices

Of course, projections and actuals may vary based on how various uses are classified and clustered, but our interest here is in identifying the major broad categories.

Production environments with repetitive work routines are at the top of the list because they can use digitalization to quantify their performance in terms of measurable outputs. Many of them already use automation and control systems to a certain extent. Adding IoT in such environments can improve the relevant key performance indicators and demonstrably justify the return on investment. Such facilities may have thousands of sensors that operate in smaller control loops to manage specific machines or production lines, using different and often proprietary data formats and protocols. IoT systems may be used to break those control and management silos and provide system-wide insights that can lead to global optimization and improved efficiency of a production site or ultimately the enterprise.

In addition to quantified insights into the facility, near-term uses and benefits of IoT in production environments include anomaly detection, preventive maintenance, and fault prediction based on combinations and correlations of sensor data.

The Internet connectivity allows a manufacturer to aggregate data from large segments or their entire installed base of products to create data-based profiles and to provide expert operational guidance to its customers and their IoT systems. It also makes it possible to track the rate and nature of defects for design improvement purposes.

Cities can benefit from the use of IoT to improve their operation and services in diverse areas, such as transportation and parking, street lighting, energy consumption, sustainability, and public health and safety. The grand vision is to provide holistic insights and coordinated actions across systems in a city. In practice, this will require instrumenting and automating of individual systems and achieving interoperability to share collected data for global insights and control. There are many projects and much experimentation with creation of smart cities that should identify the most promising uses and yield tangible benefits when fully functional systems are deployed.

Human health and fitness applications often work in conjunction with wearable devices that are capable of constant monitoring of user activities and some vital signs. This provides a range of potential uses, from tracking and encouraging exercising to detection and recording of irregularities that can led to data-informed diagnostics and personalized treatment. The potential for continuous monitoring and detection of anomalies at the moment they occur is a significant qualitative departure from the current practice of measuring of vital signs infrequently or when visiting the physician's office. Patients with chronic conditions can additionally benefit from continuous monitoring of the effectiveness of their treatment and intake of medications, as well as inform the physicians about patient's adherence to the prescribed regiment. Long-term potential impact of IoT on the practice of medicine can be truly transformational as "quantified self" provides detailed and continuous data on an individual. This in turn allows personalization of treatment based on rich and relevant data, as opposed to being guided by broad averages based on sparse and randomly sampled data in medical offices and studies, as is the norm today.

In *automotive uses*, self-driving and autonomous vehicles are probably the most publicly visible instantiations of IoT systems. They are based on extensive visual and ranging sensing of a vehicle's surroundings processed by AI-driven algorithms that produce controls and commands for autonomous navigation. Due to latency and autonomy requirements (the car should not crash if the cloud becomes unreachable), much of this processing is done locally at the edge, within the vehicle itself. However, data from perceived situations and outcomes, both favorable and unfavorable, are sent to the cloud and aggregated to refine the navigation algorithms and update the local versions executing in vehicles. Moreover, data gathered from individual vehicles may be used to continuously refine and update mapping information that is useful to drivers and essential to autonomous navigation. A car manufacturer can use data from its connected cars to get insights into typical usage patterns, type of travel and distances driven, battery usage for electric vehicles, types and rates of anomalies and failures, and the like. All of this information provides valuable guidance for product improvements and design of future models, as well as providing a basis for a variety of additional services, customer loyalty programs, and targeted offerings. One new service being considered for autonomous vehicles is to make them available for rental or ride sharing when not in use by their owners – by driving themselves to and from their service destinations.

Research and development of vehicle-to-vehicle networks is under way with the expectation that it will improve safety and efficiency by alerting nearby drivers to road hazards, such as treacherous road conditions, e.g., ice and fog, traffic accidents, and congestions. This can improve public safety and traffic management through aggregation of data about and from the vehicles moving in and out of areas of interest. Other obvious and more incremental applications of IoT technologies in automotive systems include fault prediction and detection for maintenance and operational purposes.

In retail applications, the focus tends to be on user experience, real-time tracking of user activities, and inventory for data-driven management of the supply chain.

Oil, gas, and mining are usually characterized by equipment monitoring in geographically dispersed areas, with harsh conditions and lack of access to wired infrastructure, such as electrical power and networking [10].

Similar geographically dispersed and poor infrastructure conditions exist in agriculture uses when used for monitoring of soil condition, crops, and livestock. Obviously, the applications and services are very different.

Transportation and logistics systems include a variety of applications centered around tracking goods in motion and fleets of ships and trucks that deliver them.

Utilities are expected to benefit from IoT in production and distribution via a smart grid and consumption and load monitoring via smart meters. Other major potential improvements include the ability to measure the load in real time and adjust production accordingly. Even more interesting is the possibility to sculpt the load, i.e., activate or delay schedulable consumers – such as EV chargers and appliances – to coincide with the production cycles of renewable but variable energy sources, such as wind and solar. Adaptive matching of supply and demand by means of load shaping is useful because storage of energy is not yet commercially cost-effective and mismatches result in waste or outages.

In *homes* the current focus tends to be on automation of chores and maintaining user convenience and comfort through home automation for security and ambient control such as lighting and temperature, smart thermostats, and smart appliances. Voice assistants that can access the Internet and control some home devices are becoming popular in some markets. They can act as input, output, and control point or gateway for other home devices. Unfortunately, the use of different standards leads to fragmentation and general inability to control home devices from different manufacturers as a collection with coordinated behaviors to improve user experience.

In general, changes resulting from introducing IoT capabilities can be (1) transformational changes that improve the effectiveness of a system and (2) new uses and applications that may enable new business models. IoT uses and applications described earlier combine components of the two to various degrees.

One of the features of IoT systems that enables some interesting new service and business models is the ability to continuously and remotely monitor the physical world. Coupled with the Internet connectivity, it is giving rise to what is often referred to as "something as a service" model. One of its incarnations consists of providing the physical equipment with monitoring and control on an ongoing service basis, instead of an outright purchase. For example, in an engine or power as a service scenario, aircraft engine manufacturers may provide engines to airlines and offer continuous monitoring of their operational performance in exchange for service fees that may include components based on usage, such as operating hours or segments flown. For the manufacturer, benefits can include the constant recurring service revenue flow and the real-time insights into their products in operation across the customer base. Monitoring data can be used for early detection of anomalies and to improve maintenance. Aggregations of data can be used to improve operational analytics and AI algorithms. In addition, they can provide insights and learnings to guide product improvements and future evolution. For airlines, benefits can include expert monitoring and maintenance guidance, as well as conversion of the up-front capital expenses for engine acquisition into operational expenses that can be covered from the ongoing revenue stream. This can reduce the need for capital loans and looks much better on the corporate performance sheet.

The presented examples illustrate that the possibilities brought by the IoT systems and by introducing the real-world dimension and awareness to the Internet have significant commercial and transformational potential. However, we are very early in the development and deployment cycles of IoT technology, and – if history of computer applications is any guide – its true potential remains to be realized by the future applications yet to be invented.

Issues and Challenges

IoT system implementations and their applications face a number of challenges that need to be addressed in order for the full potential of IoT can be realized. The major ones include:

- Technology and standards
- Security and privacy
- Business value and adoption

Technological challenges include a system design that is suited to the operational characteristics of IoT systems and addresses its differences from the traditional Internet. Predominant among these is the need to achieve interoperability of IoT data across specifications and domains. The current fragmentation of standards and proprietary approaches create silos that limit usefulness and the ability to harness the benefits of big and diverse aggregations of IoT data. One major analyst study [4] estimates that achieving interoperability can increase the potential IoT business impact by 40%. In its absence, a great opportunity remains in danger of not being realized.

Security is important in all computer uses and especially in enterprise and production environments where breaches can result in significant financial and reputational losses. It is even more important in IoT systems that can directly impact the physical world where cyberattacks can cause not only major damage and disruption but even endanger the safety of people and the environment. The problem tends to be further aggravated in IoT systems that are connected to legacy operational systems, such as utilities and manufacturing, that may have inadequate and ossified security protections. In the consumer space, individual devices and home installations are often insecure and may create security and privacy exposures due to device design and configuration deficiencies resulting in inadequate security or one that deteriorates over time due to lack of updates and ongoing security management.

Privacy and the ability to safeguard user data are also of major importance in IoT systems, since their data streams can provide considerable details and real-time information on user activities and location. Unchecked revelation of such data may be exploited for not only unwelcome and potentially intrusive surveillance but also for nefarious purposes such as break-ins or theft when users are known to be away.

Finally, large-scale commercial success and wide deployment of IoT in industry depend on creating uses, applications, and services that create value. While experimentation with IoT and proofs of concepts abound in industry, one of the key challenges is to demonstrate tangible business value. This means identifying applications with considerable revenue potential that produce positive business outcomes and measurable improvements, such as increased efficiency, user experience, reduced costs, or some combinations of those.

IoT Systems: A Reconnaissance Flyover

This section provides a brief overview of the entire IoT system in order to identify key functions and components and how they relate to each other. The presentation is provided from the two different points of view to highlight the purpose and structure of various parts of IoT systems: (1) functional view and (2) infrastructure view. The functional view focuses on what needs to be done, i.e., key processing stages of IoT data on their way from capture to processing and actuation in some form. This is followed by the infrastructure view that focuses on where and how things get done, i.e., the key infrastructure and hardware components that execute those functions.

IoT System Functional View

The primary purpose of an IoT system is to collect real-world data and make them available to services and applications in order to create insights and act upon them by affecting the real world in some way [11]. Implementation of those functions requires an infrastructure to run them and control functions to keep them secure and operational. In this section, we focus on the functional aspects and cover other views in the subsequent ones.

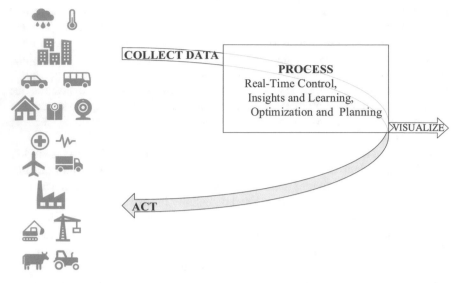

Fig. 1.2 IoT system functional view

Figure 1.2 depicts a highly abstracted functional view of an IoT system with focus on data flows and types of processing from capture to output actions. It highlights the key stages in IoT data and control flows: (1) data collection, (2) processing, and (3) acting upon the world based on the outcomes.

Data Collection

Data collection starts at the edge, with a sensor acting as a physical-cyber interface that monitors and reports states of some physical entity or device. The intent is to produce a digital representation suitable for use in the cyberspace. This process may involve many practical details, such as signal conditioning, analog-to-digital conversion, scaling, and conversion to engineering units for subsequent processing. These operations are important for the implementation of IoT systems, and we cover many of them in subsequent sections and chapters. However, from the functional point of view, data collection results in digitized state samples of the physical world suitable for processing by the applications and services in the cyber domain.

Relatively early in the process, metadata needs to be captured and used to annotate the data. In IoT systems, metadata generally describe the nature and context of data capture, such as the sensor type, its location, and in some cases structural relationships to other elements of the system.

Data Processing and Visualization

Types of IoT data processing range from the simple control loop algorithms performed on the incoming streaming data as they arrive to the sophisticated forms of analytics and machine-learning algorithms that operate on combinations of streaming and archived data, events, and records of past behaviors and observations of the system.

Various steps of data processing may be implemented in increments or in entirety in the different components of IoT systems. In general, their scope and complexity tend to increase in the higher levels of system hierarchy, where more processing, power, storage, and larger aggregations of data are available.

Common data preprocessing steps may include filtering, aggregation, and comparisons to detect if the sampled data are in a special condition that warrants additional action, such as creation of an event or notification. When detected, they are forwarded to the processing services that may include some combinations of event-driven control loop algorithms, user notifications, or operator alarms.

The next level of sophistication in data processing is to provide optimizations and predictions of system behavior based on its current state, past behaviors, and guidance form algorithms, such as analytics and machine-learning models. Subsequent stages consist of learning system behaviors in order to transform that knowledge into effective insights and control actions.

In industrial and complex control systems, it is customary to visualize the system state and points of interest to system operators. These visualizations normally include data on the vital indicators of system state and notifications and alarms when faults or anomalous behaviors are detected. Operators usually have the option to zoom in and inspect any data point of interest for information and analysis purposes.

Improvements due to IoT deployments usually happen in stages. The main phases typically progress through the following stages: (1) data collection and visualization, (2) insights and learning, and (3) optimizations and actions.

The first stage is to instrument the system under observation by installing and connecting sensors and devices with the rest of the IoT system. The resulting monitoring of collected data leads to insights and evidence-based understanding of the observed systems. While algorithmic analysis is often the goal, a somewhat unappreciated but important aspect of IoT instrumentation and continuous data collection is that they can immediately provide valuable insights to system operators. Human operators familiar with the system can use their experience and natural intelligence and act as very good "inference engines" in generating insights that they can deploy for more effective management of system operations. Moreover, they can also identify areas of potential improvements for analytics to focus on.

The third stage is to deploy the predictive and prescriptive forms of analytics and AI algorithms, informed by the prior human and machine insights and learnings.

Acting

Acting upon insights and predictions is the output and the ultimate purpose of deploying IoT systems. Actions can take different forms, from simple remote actuation initiated by operators in response to visualized conditions in a basic monitoring configuration to automated guidance of control points that proactively manages conditions in a smart building to maximize user comfort and optimize energy efficiency. Actions can be implemented as direct actuation or indirectly, in the form of advice to system operators or optimizations resulting in adjustments to the manufacturing process. They can also include identification of cause of failures and anomalous conditions followed by direct or indirect execution of the appropriate remediation actions.

With the addition of interoperable data formats and Internet connectivity, the scope of data aggregation in IoT systems can grow to potentially any level including multi-domain systems such as smart cities and regions. Such data may be used for detecting and acting upon regional and even global insights. For example, one could aggregate behavioral and energy efficiency data from all IoT-managed buildings in a large region and use it to train and improve the AI and ML algorithms for cross-domain effectiveness in all of them.

IoT System: Infrastructure View

This section covers the IoT system infrastructure components, including data processing, storage, and communication, that host and execute its functions outlined in the previous section.

Large IoT installations can be complex distributed systems with a variety of components and multiple levels of hierarchy. Figure 1.3 illustrates some key infra-structure components of an IoT system. Edge components are depicted towards the bottom, the communications layer mostly in the middle, and upper levels of system hierarchy ending with the cloud are shown on top.

Edge components include sensors, smart things, gateways, and fog nodes. The communications layer provides connectivity among system components that they can use for horizontal peer-to-peer interactions within a level of system hierarchy or for cross-level communications towards the Internet and the cloud. The cloud portion depicts the back-end part of the IoT infrastructure where large-scale data aggregation and processing take place.

Edge Components

At the edge layer, Fig. 1.3 depicts several types of components, including smart things and gateways. Smart things are generally characterized as devices that con-tain sensors or actuators and contain sufficient functionality to perform some local

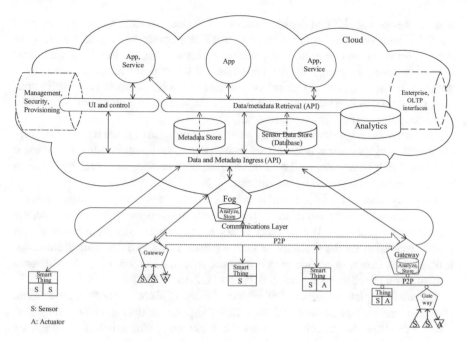

Fig. 1.3 IoT system infrastructure view

operations, connect to the Internet to report their data or receive actuation requests, and interact with services. Smart things in Fig. 1.3 are shown as connecting directly to the cloud, to a fog node as a processing intermediary, and to other network nodes, such as gateways, in peer-to-peer configurations.

In practice, communication is carried at the behest of software agents that implement the relevant functions and services at the sending and the receiving party. If both parties reside on the same network or are at the same level in system hierarchy, this communication is referred to as peer-to-peer (P2P) and sometimes more generally as machine-to-machine (M2M) communications. Technically, all communications between IoT devices themselves as well as with the applications and services are of the M2M type, so we use the term P2P to depict peer-level communications, which are shown as horizontal paths in Fig. 1.3, as opposed to the edge to cloud which would follow the vertical paths. Depending on the underlying physical network, a single node may be able to engage in both types of connections.

All of these directions are conceptual in the sense of indicating the hierarchical position of endpoints in a given communication. With the Internet as the underlying network layer, any addressable point can connect to any other addressable point anywhere in the world, subject only to authorization.

P2P communications can be somewhat distinct in the sense that they can take place between nodes using simplified or application-specific protocols that are not Internet compatible, such as Bluetooth or legacy industrial devices. This can reduce the load and simplify design of the constrained nodes. This is illustrated in the lower right corner of Fig. 1.3 that depicts a cluster of devices, a thing, and a gateway connected to each other and another gateway in a P2P manner. As long as they support a common set of private protocols, the three nodes can engage in P2P communications. However, to qualify as IoT endpoints, at some point in the system, say a gateway, a transition to the Internet compatible protocols needs to be made in both directions, so that functional exchanges of messages can take place with the authorized endpoints anywhere on the Internet.

Gateways are edge devices in IoT systems to which one or more basic sensors are connected and dependent upon for wide-area connectivity and optional additional services. Their basic role, as the name implies, is to provide connectivity between locally connected sensors and the Internet.

Gateways can reside in physical structures, such as a building, where an Internet access point already exists and the gateway needs to be connected, for example, via an Ethernet connection. If a gateway is not near a usable existing Internet access point, depending on the distance to such a point, it may provide a medium-range link, such as LoRa, or a wide-area link to the Internet, say via a private wide-area network or a telephone company link such as GPRS or LTE.

The early implementation of IoT systems tended to place most of the processing, analytics, and storage in the cloud. In such settings sensor data are sent to the cloud directly or via predominantly communications gateways with minimal intermediate processing. As discussed in detail in chapter "The Edge", this approach has drawbacks that include longer communication and processing latencies, potentially high bandwidth usage, and unavailability of control functions during communications and cloud outages. This can be a problem in industrial applications that may require fast response times and a certain degree of local autonomy, including the ability to provide at least the basic control at all times regardless of cloud availability. Placing of processing and storage functions closer to the edge tends to alleviate these problems.

Data processing functions can take place at more powerful gateways or at some variants of fog nodes that can perform some analysis and storage functions, as illustrated in Fig. 1.3. In practice, the term fog tends to be used for relatively higher-powered edge nodes that are used for processing and storage and may or may not provide communications gateway functions and interface to sensors. To illustrate this point, Fig. 1.3 shows one potential placement of a fog node somewhere within the communications layer and possibly touching or residing in the cloud. Fog nodes are supposed to bring cloud-style processing and considerable power closer to the edge. The name is a metaphoric allusion to fog as the part of cloud that is close to the ground.

Rather than conceptualizing processing and storage nodes as being of a given type, placement of data processing in IoT systems should be viewed as a functional continuum that can span the range from edge nodes all the way to the cloud. As discussed in subsequent chapters, IoT systems should be implemented to allow flexible placement and late mapping of functions to hardware. This allows allocation of functions to meet the needs of a particular system and to even respond to load changes dynamically by reallocating them at runtime if needed.

Communications Layer

The communications layer enables a vast array of edge devices and things to exchange messages with each other, the rest of the IoT system, and ultimately the Internet. In IoT systems, the communications layer may include a variety of wireless and wired links, spanning local areas and including long-haul connections, such as wide-area networks and telco LTE variants. It may represent a complex infrastructure of links, bridges, and routers that can transport payloads from local point-to-point segments all the way to any endpoint and application on the Internet. In addition to hardware, this requires implementation of a number of network layers and protocols which in IoT systems are commonly based on the Internet blueprint as discussed in chapter "Communications". Moreover, whenever possible IoT nodes should make use of the already installed Internet infrastructure with its global reach and numerous access points.

The choice of a specific type of IoT physical connectivity can be a complex decision involving considerations such as system topology, range limits of various wireless technologies, location of usable Internet access points, bandwidth requirements, and costs of communication links in terms of initial interfaces and setup as well as the ongoing operational and bandwidth expenses thereafter. Another consideration in wired system is the cost of acquiring and laying out wires from sensors to gateways, which can influence the number and placement of gateways in the field.

Cloud

The cloud is the ultimate meeting place of large-scale IoT data, applications, and services that operate on them. Streaming and archived data are made available to authorized applications and services via APIs and queries. Cloud implementations generally provide the top end data aggregations and the server, storage, and connectivity infrastructure to execute services and applications that use them.

In terms of implementation, the term cloud generally refers to the large collections of servers and storage located in data centers that can be accessed via the Internet and allocated to services and applications in a dynamic and elastic manner. The details of cloud operation in IoT systems are discussed in chapter "Cloud".

IoT data coming to the cloud may be processed in-flight as they arrive, stored for subsequent processing, or both. System-level rules and policies in effect determine how individual data should be processed and routed to the appropriate destination points and services.

The cloud portion of Fig. 1.3 also illustrates the logical functions of aggregation and storage of sensor data and metadata. As discussed later, in the actual implementations, they may be kept as separate or combined into a single database.

Analytics and ML algorithms may be applied to create insights and actionable recommendations by processing data at various points in an IoT system. As discussed in chapter "Cloud", such algorithms tend to be developed in the cloud where they can use the massive processing resources and large data aggregations, both of which are required for algorithm training and producing of system-wide insights. When completed, ML algorithms can be deployed at other nodes in the system, such as the edge, to perform inferences and spot trends using locally available data. Figure 1.3 also shows a separate storage for the analytics, which should be regarded as a logical function indicating the need for storing of derived data and algorithms.

An optional data-related function shown in the cloud in Fig. 1.3 is the interface to enterprise systems and online transaction processing systems (OLTPS). In various installations, enterprise systems may include any combination of systems such as ERP (enterprise resource planning), CRM (customer relationship management), and variants of e-commerce. In general, it is useful to interface them to the IoT portion for data exchanges and to facilitate holistic insights that can benefit all sides of operation of a complex system.

Figure 1.3 also highlights two important interface points in an IoT system, notably the data and metadata APIs between the edge components and the cloud and the data retrieval APIs for live streams and stored data used by the applications and services executing in the cloud. In general, APIs should enable applications to query, search, and access data and metadata of interest. Formalizing the types of interactions and data formats that they support is not only a good design practice; it also provides the foundation for modularity and interoperability in implementation of IoT systems.

Control Plane

The components and functions described so far implement the primary objective of an IoT system to collect, process, and act on data. Parts of the system that carry out these production activities and implement the related system flows are commonly referred to as the *data plane* or user plane. The task of keeping the IoT infrastructure itself running and secure is usually delegated to a separate system overlay that is referred to as the *control plane*. It is partially depicted in Fig. 1.3 as the service and database labeled security, management, and provisioning. Although seemingly playing a supporting role to the primary mission, security and management are essential to keeping an IoT system up and running securely and with integrity so that it can fulfill its intended purpose and not be a threat to safety of the people and the environment.

During normal operation, control plane systems constantly monitor activities that may impact security and availability. This is accomplished through the network of management agents that are installed on nodes and system components to observe and report status and changes. Their reports are customarily aggregated and visualized to operators at a central control point. The agents are also used to distribute and manage security policies and credentials, change configuration, and update firmware and software as necessary.

An important function of central monitoring is the detection and analysis of suspicious behaviors that may indicate probes from the adversaries or breaches of security. When incidents are detected, the handling mechanisms and policies are activated to mitigate the situation by identifying and isolating the compromised parts of the system. Details of those operations are presented in chapter "Security and Management".

While the system is in operation, new nodes may have to be added, existing ones patched, and old ones decommissioned without bringing down other parts of the system or relaxing its security posture. Security and management systems are involved in preparing nodes for joining the system in the early stages of their lifecycle that precede activation. During the process of node commissioning and provisioning that follows installation, they are issued system identities and security credentials necessary for authentication and for secure operation upon their activation in the system. During that time, nodes are also entered into device registries and other back-end systems that may need to be involved in their operation, such as the billing, asset management, and support.

Book Organization

The rest of this book is organized as follows. Chapters 2 through 5 cover in detail the concepts and design considerations of system elements and components.

Chapter 2 covers by starting with sensor data acquisition and processing and continuing to the edge functionality that includes event processing, storage, local control and scripting, and interfacing to sensors and actuators as well as to the external communications and the cloud. After discussing the trade-offs involved in the functional placement of components in distributed systems on the edge-to-cloud continuum, including the fog, the chapter continues with a description of hardware and software considerations involved in the edge-node design. It concludes with a brief description of the architecture and implementation of an open-source edge framework as a practical instantiation of the concepts covered earlier in the chapter.

Chapter 3 describes a layered network design which is the underpinning of the Internet and a useful blueprint for the IoT system design. It continues with the coverage of wireless and constrained networks at the edge, including the IEEE 802.15.4 and its derivatives and variants, followed by the description of 6 LoWPAN bridging to the Internet, and some non-IP-based networks that have a large base of installed devices, such as the ZigBee and Bluetooth. Subsequent sections cover cellular offerings in the licensed spectrum, including their IoT adaptations, such as the NB-IoT. The chapter concludes with the exposition of the Constrained Application Protocol (CoAP) which is commonly used in IoT systems as a lighter-weight functional substitute for the HTTP and the popular messaging and queuing implementation of the IoT publish-subscribe mechanism MQTT.

Chapter 4 focuses on key elements and functions of IoT cloud core components, including data ingestion via edge-cloud gateways, in-flight stream processing, and short-term and long-term storage systems suitable for IoT applications. The second half of the chapter focuses on analytics and optimization algorithms. It starts with a real-life example and side-by-side comparison of the effectiveness and use of the algorithmic machine analytics and the traditional optimization methods. The rest of the chapter covers principles of machine learning, operation of artificial neurons and networks, and types and uses of ML systems. The closing section discusses the process and details of creating and operating a ML model, including model selection, training, and optimization.

Chapter 5 covers the control plane, security, and management systems. It covers types of security threats and attacks in IoT and OT systems, followed by the security planning and analysis steps that include risk assessment and threat modeling that indicate what should be guarded against and which mitigation techniques to use. A section on cryptography overviews key foundational elements of security design, including symmetric and public-key cryptography, key exchange, message authentication, and digital signatures. This is followed by the treatment of endpoint security, including hardware security modules, such as TPM and TEE that facilitate secure booting and software execution, followed by the software isolation mechanisms, such as virtualization and containers. The section on network security covers transport-level security and network isolation and segmentation techniques. This is followed by a treatment of security monitoring, incident handling, and systems management in major stages of node lifecycle, including provisioning and activation. The last two sections cover privacy and a summary putting it all together section.

Chapter 6 focuses on IoT data formats and representation. It explains why IoT data representation needs to include semantic annotation which makes it different from the primarily textual World Wide Web. The common structure of IoT information models is described, including object types, attributes, interactions, and links. A separate section covers IoT interoperability among nodes through the use of shared information models. This is followed by the description of data serialization and format of payloads as they are exchanged on the wire. A section on metadata covers its definition and importance and provides examples of use and benefits. IoT frameworks are introduced as one approach to achieving functional interoperability among compliant nodes within a domain. Cross-domain interoperability is covered in a separate section. It identifies three levels of interoperability and focuses on the value and need for interoperability across domains and in large data aggregations that are a prerequisite for holistic system management and AI.

Chapter 7 covers IoT data standards and highlights the salient features of several of them with a focus on data information models and interoperability. It illustrates the scope and directions of the ongoing standardization work and reviews some of the more influential efforts including IPSO, OCF, WoT, Haystack, and OPC UA. Examples of their definitions of a basic sensor type are included for comparison. The chapter concludes with a summary of the commonalities, differences, and some limitations of the presented approaches to the problem.

Chapter 8 provides an overview of several major commercial IoT platforms to illustrate the scope of what is available to IoT system designers as potential building blocks. It also points out the structural similarities in the architectures of the presented systems and differences in their scope and emphasis.

Chapter 9 summarizes design and integration considerations involved in putting together an entire IoT system. It is intended to serve as a high-level checklist and an expansion of some of the issues and recommendations provided in the prior chapters on system components. It also includes a detailed example of the design of an actual system, starting with a definition of purpose and specification and design outline, followed by the implementation, experimental results, and a pilot evaluation with the actual users. It outlines key stages in the system development from the inception to completion and includes a discussion of redirections and changes made based on the insights gained in the design and implementation process.

References

1. Gabbai, A. (2015) 'Kevin Ashton describes the internet of things' *Smithsonian Magazine* [Online] Available at: https://www.smithsonianmag.com/innovation/kevin-ashton-describes-the-internet-of-things-180953749/ (Accessed Dec 15, 2019)
2. McKinsey Global Institute (2015) 'The internet of things: mapping the value beyond the hype' [Online] Available at: https://www.mckinsey.com/~/media/McKinsey/Industries/Technology%20Media%20and%20Telecommunications/High%20Tech/Our%20Insights/The%20Internet%20of%20Things%20The%20value%20of%20digitizing%20the%20physical%20world/The-Internet-of-things-Mapping-the-value-beyond-the-hype.ashx (Accessed Dec 15, 2019)
3. ITU-T Y.4000/Y.2060 (2012) 'Overview of the internet of things' [Online] Available at: https://www.itu.int/ITU-T/recommendations/rec.aspx?rec=y.2060 (Accessed Dec 15, 2019)
4. Wikipedia 'Internet of things' [Online] Available at: https://en.wikipedia.org/wiki/Internet of_things (Accessed Nov15, 2019)
5. Maker.io [Online] Available at: https://www.digikey.com/en/maker/ (Accessed Nov 15, 2019)
6. Makezine [Online] Available at: https://makezine.com/ (Accessed Dec 15, 2019)
7. Raspberry Pi [Online] Available at: https://www.raspberrypi.org/ (Accessed Nov 15, 2019)
8. Arduino [Online] Available at: https://www.arduino.cc/ (Accessed Dec 15, 2019)
9. SeeedStudio [Online] Available at: https://www.seeedstudio.com/ (Accessed Nov 15, 2019)
10. Hanes, D or Barton P.? et al. (2017) *IoT fundamentals: networking, technologies, protocols and use cases for the internet of things*, Cisco Press, Indianapolis, IN.
11. Chou, T. (2016) *Precision: principles, practices and solutions for the internet of things.* Cloudbook Publishing, USA.

Chapter 2
Edge

An IoT system edge is a touchpoint and interface between the physical world and the cyber world. On the input side, the IoT edge may be functionally viewed as an endpoint that can capture real-world data to make it available to authorized clients anywhere on the Internet. In principle, this implies that the endpoint should be addressable and capable of communicating the data in a format understood by the rest of the IoT system in which it operates. In practice, some of the edge nodes may not have the processing power or the Internet Protocol (IP) connectivity to perform these functions, and they rely on other intermediate nodes in the data path, such as gateways, to provide the missing pieces.

Edge nodes and things span a fairly large spectrum of device types, ranging from integrated smart sensors, connected devices, and things that include sensors, gateways, and fog nodes that can provide sophisticated functions to groups of sensors and things connected to them.

The primary function of an edge node is to acquire sensor data and transmit them to clients as appropriate. This may involve a number of preprocessing and scaling steps to convert raw sensor data into engineering units and in a format suitable for use by the rest of the system. Optional data processing steps at the edge include filtering, generation of events and alerts, local control, and storage. Advanced functions include edge analytics that often operates in conjunction with the back-end AI and ML algorithms.

The functions described above implement the production part of IoT edge node operations. Edge nodes can be complex instantiations of computer systems, and they need to be managed and secured as such. The edge is also a boundary between physical things with sensors and the secure and managed IoT infrastructure that hosts and executes applications and services. This may require additional provisions for and an extended role in edge-node security as described later in this book.

In the remainder of this chapter, we describe most of these functions in detail and how they are typically combined to implement edge nodes.

© Springer Nature Switzerland AG 2020
M. Milenkovic, *Internet of Things: Concepts and System Design*,
https://doi.org/10.1007/978-3-030-41346-0_2

Sensors and Actuators

A sensor is a device that detects some measurable aspect of the physical world state (stimulus) and converts it to a processable output, commonly an electrical signal. Sensor-produced signals are representations of monitored or measured physical phenomena, and they come in analog and digital formats. There are hundreds of analog sensor types that can measure a variety of things including pressure, temperature, humidity, light intensity, air quality, presence, audio, video, touch, position, motion, orientation, magnetic fields, gas and liquid flows, liquid level, voltage, and current. Digital states of the real world are exemplified by devices that can assume a relatively small number of discrete states, such as a valve being open or closed, or a light switch that is on or off.

In general, the definition of IoT sensing should also include human inputs and observations on the physical world, such as that a street is flooded or that a public trash receptacle is full. In IoT systems, such inputs are usually integrated and processed in the higher levels of system hierarchy, such as the cloud. Consequently, we defer their treatment for later chapters.

Sensors and devices that contain them vary widely in terms of their functions, processing power, and connectivity. They range from basic sensors that provide raw signals from transducers to fairly complex devices, such as smart thermostats with multiple sensors, actuators, and set points with a direct connection to the Internet. This section covers the details of acquiring and preprocessing of sensor data so that they can be used by applications, services, and devices in the rest of an IoT system.

Sensor Signal Processing

This section describes the steps involved in acquisition and preprocessing of sensor data. It basically involves the following steps:

- Sensor (transducer) produces measurement as electrical signal
- Signal conditioning, transfer, and conversion to digital
- Conversion of digital values to engineering units

Signals from raw sensors and transducers need to be brought to the appropriate system interface point where they can be acquired and converted to digital representation for further processing. Raw transducer outputs typically come in the form of voltage or current signals that vary in some proportion to the stimulus, i.e., the measured physical phenomenon. Those signals may need to be driven through some signal conditioning and amplification circuitry to match the range of their terminating points and interfaces, such as 0–5 V for voltage or 4–20 mA for current inputs. For digital signal inputs, this is often a conversion to 0 or 1 value in the corresponding input register that can be read programmatically by the receiving entity. For analog signals, an additional analog-to-digital converter (ADC) circuit is needed to sample and convert the input signal into its corresponding digital value. Digitized values are then post processed to convert them to engineering units for use by the rest of the IoT system. Ultimately, the accuracy of the result depends on a number of factors, including the accuracy of the sensor itself and of the components involved in its conversion path.

Digitization of analog values produces a so-called *quantization error* as continuous analog values are represented by a limited range of digital numbers. This is essentially a roundoff error. Its magnitude depends on the ADC resolution defined by its number of bits. Another factor in play is the accuracy of the ADC itself, i.e., how many bits in the digital output code represent useful information. It is a function of the internal circuitry and noise from the external sources connected to the ADC input.

Acquisition of the fast-changing input signals imposes a choice of the sampling frequency, i.e., the rate of taking measurements. This is necessary in order to be able to reconstruct the original signal, subject to sensor accuracy and digitization errors discussed above. According to the Nyquist sampling theorem [1], the input signal must be sampled with at least twice the highest frequency that needs to be measured. For example, audio signals are typically sampled at 44,000 Hz to be able to reproduce output signals accurately within the 20–20,000 Hz, which is generally regarded as the range of audible frequencies for humans. Insufficient sampling rates result in inability to reconstruct the original signal veritably, and they are referred to as *aliasing errors*.

Getting into the details of processing raw sensor data in general is beyond the scope of this book. It is the subject of signal processing and control systems theory. The intent here is to point out there may be quite a few transformations and computations involved in the process. Data acquisition and the appropriate signal processing logic need to be implemented in the path from the sensor to its data output software interface on the IoT gateway. More sophisticated and smart sensors may perform these actions on the integrated controller to produce a linearized and compensated output value. For sensors providing raw data from the transducer, these steps may be performed in the input driver of the node that it is connected to or a part of.

IoT system implementers need to be aware of and prepared to tackle these issues or opt to choose sensor types and assemblies that contain internal linearization and conversion logic that provide the necessary functions.

An Example: Thermistor

The remainder of this section describes some of the details involved in sensor data acquisition, processing, and conversion to engineering units by means of an example. The intent is to illustrate some of the intricacies and considerations that may be involved in producing a usable sensor reading. To make it realistic, we use an example of a commercial sensor, a thermistor in this case. Readers not interested in the details of processing of raw sensor data may omit this section without loss of continuity.

Thermistors are popular transducers used to measure temperature. They generally give repeatable results and are relatively inexpensive. Thermistors are used to measure temperature in air conditioning systems, home appliances, and 3D printers, and a number of them may be found in a modern automobile.

Thermistors have the property of changing their resistance with temperature changes, hence the name *thermally sensitive resistors*. They are made of ceramic materials produced by heating various metal oxides. A common type of thermistor has the property of reducing its resistance as temperature increases and is called the negative temperature coefficient (NTC) thermistor [2].

Since voltage is easier to measure the resistance, thermistors are usually combined with a few other components to produce voltage that varies in some relationship with the temperature. Figure 2.1 illustrates a simple voltage divider where a thermistor, denoted by R_T, is combined in a series with a fixed resistor and the line is fed by the reference supply voltage, V_{CC}. Fluctuations of thermistor resistance with the temperature of the medium where it resides, such as coolant or air, cause corresponding fluctuations in the output voltage of the circuit, denoted by V_T.

In the configuration shown in Fig. 2.1, sensed voltage can be expressed as:

$$V_T = V_{CC} \times \left(R_T / \left(R_T + R_B \right) \right)$$

Thus, R_T can be computed as

$$R_T = \left(R_B \times V_T \right) / \left(V_{CC} - V_T \right))$$

where R_T is the variable thermistor resistance, V_T is the produced (measured) variable voltage, R_B is the fixed bias resistor, and V_{CC} is the reference voltage supplied to the circuit. In general, R_B should be chosen so as to provide a reasonably wide response over the temperature range where the thermistor will be used.

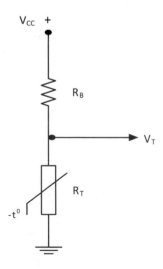

Fig. 2.1 Thermistor circuit

The thermistor variation of resistance in relation to temperature tends to be non-linear. The curve for the one used in our sample fuel injector is shown in Fig. 2.2. As indicated, the relationship is nonlinear, sloping downward since this is an NTP thermistor. The shape and values on the curve are device specific, and they can be found in the manufacturer's data sheet.

In this particular device [2], the thermistor is configured as shown in Fig. 2.1 with the bias resistor R_B of 2490 Ω (Ohm) and the V_{CC} of 5 V. For example, at $-20\,°C$, our sample thermistor resistance is 16,150 Ω, and at 97 °C, it is 199 Ω. Voltage produced by the circuit depicted in Fig. 2.1 in those two cases may be computed as

$$V_T = 5 \times \left(16150 / \left(16150 + 2490\right)\right) \cong 4.332 \ V$$

and

$$V_T = 5 \times \left(199 / \left(199 + 2490\right)\right) \cong 0.370 \ V$$

respectively.

We can measure the output voltage V_T, compute the resistance R_T, and use the data from the curve in Fig. 2.2 to determine the temperature. The output voltage from the thermistor circuit would typically be driven to the input of an ADC that would provide a corresponding digital reading. Depending on the distance between the two, some additional signal amplification and conditioning circuitry, such as an operational amplifier, may also be included.

An ADC usually produces integers corresponding to the range determined by its number of bits. A 10-bit ADC provides output values in the range of 0–1023. For a 5 V input range, that means that it can measure voltage in increments of approximately 0.00488 V. In our example, values produced by the ADC would be readings of approximately 888 for 4.332 V and 76 for 0.370 V. The latter corresponds to 199 Ω and thus represents the temperature of 97 °C.

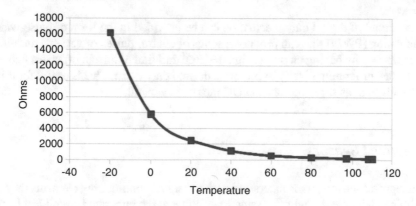

Fig. 2.2 Sample thermistor temperature resistance curve [3]

When a computational formula is not readily available or easy to compute on smaller microcontrollers, nonlinear data are typically stored in the form of a lookup table consisting of pairs of inputs, one being the measured value and the other its corresponding output. In our example, the lookup table might consist of resistance-temperature pairs (Ω, °C) and include entries (16,150, −20) and (199, 97) for our two sample resistance and temperature values. Or a more expedient lookup table may be constructed for our specific thermistor circuit, pairing the 10-digit binary voltage readings with the corresponding temperatures, including entries (888, −20) and (76, 97) for our example calculations. Lookup tables are fast to use at runtime but require memory to store the complete set of possible entries, up to 1024 in our example. Since 888 is the maximum value of the rated range of our thermistor, detection of higher values close to 1000 may be used to indicate a fault, such as a short circuit.

Actuators

Actuators are an output part of the IoT system that performs direct actions upon the physical world. Actuation provides the means to implement control actions as and when determined by the algorithms or system operators. Actuators can produce analog or digital outputs. Digital actuators are usually managed by writing values to actuator output registers that, past the appropriate signal conditioning and adaptation circuitry, can trigger relays or change the state of a thing, such as a power switch which can result in turning a light on or off. They can produce a single on/off action or generate timed control pulses individually or in series.

Analog actuators typically produce continuous signals that can be used to drive devices such as controlling the speed of a motor or producing sound in your headphones. In order to produce analog output signals, corresponding output data (numbers) are fed into a digital-to-analog (DAC) converter circuitry. A DAC converts a digital value into a corresponding analog output signal, basically acting as an ADC in reverse.

A variant of digital output actuation can be provided in the form of pulse-width modulation (PWM) that can produce a series of control pulses of variable duration. They operate by having an output signal whose duration of on and off states may be controlled programmatically to achieve a desired duty cycle. This is useful for driving loads such as servo motors and DC motors in general.

Integrated Sensors

In integrated sensors, transducers and optional signal conditioning elements may be combined with the digital processing logic either at the integrated circuit level or in various forms of system-in-a-package (SiP) configurations. They usually produce digitized outputs that are communicated via wired serial buses, such as I2C or SPI, to reduce the number of signal traces especially for natively analog measurements. Stand-alone sensor assemblies may also use wireless communication to their neighbors and/or communications gateways. Those obviously require additional components, such as radios, modems, and communication software, for link management and protocol processing. For example, the popular personal activity monitors contain a few sensors – such as gyroscope, accelerometer, compass, and altimeter – and use a wireless connection, such as Bluetooth Low Energy (BLE), to connect to a nearby gateway, usually a smart phone.

Some sensors, such as smart home thermostats, may contain the additional ability to connect to the Internet, often via home Wi-Fi connection. They usually connect directly to their service providers and tend to use proprietary protocols and controls not suitable for general use. As a design choice, smart sensors can support standard protocols and connect to any suitable level of hierarchy in IoT systems.

Interfacing Sensors

Signals from raw sensors and transducers, such as the thermistor presented earlier, need to be brought to the appropriate system interface point where they can be acquired and converted to a digital representation for further processing. Raw transducer outputs typically come in the form of voltage or current signals that vary in some proportion to the stimulus, i.e., the measured physical phenomenon. Those signals may need to be driven through some signal conditioning and amplification circuitry to match the range of their terminating points and interfaces, such as 0–5 V for voltage or 4–20 mA for current inputs. For digital signal inputs, this is often a conversion to 0 or 1 value in the corresponding input register that can be read programmatically by the receiving entity. For analog signals, an additional analog-to-digital converter (ADC) circuit is needed to sample and convert the input signal into its corresponding digital value. Digitized values are then post processed to convert them to engineering units for use by the rest of the IoT system.

Two of the more common ways for interfacing sensors are described in the sections that follow.

Interfacing Sensors in Industrial Automation

In industrial automation, there is a well-established practice of connecting, often through wiring, sensing, and actuating points for process control and factory-floor machinery. The automation industry [4] has largely standardized on the rail-mounted sensor and control components with the adjoining cable conduits, with power supply and operating voltages commonly in the range of 12–24 V DC and AC.

As a result, there is a wide choice of commercially available input and output devices, including sensors, relays, motor, and position controls. Standardized signal levels and paths, such as the 4–20 mA current loops, are used to provide mixing and matching of different devices and connections to system interface and control points. They may also be provided as already digitized signals interfaced via serial I2C (Inter-Integrated Circuit) or SPI (Serial Peripheral Interface) buses. Sensors and actuators may be connected to programmable logic controllers (PLCs) via their I/O ports and modules for analog and digital signals. Such modules often include signal conditioning, level adjustment, provisions for structured industrial wiring and mounting, and safety features such as isolation, grounding, and protection from input overshoots and short circuits. Industrial components are typically designed to withstand operation in harsh environments that may include high temperatures, vibration, and dust. Some of them are even designed to operate in environments that can be explosive.

Commercial offerings consist of a wide variety of sensors and actuators that use structured wiring to connect industrial sensors and actuators to the first tier of digitized control, usually (PLCs). A PLC [5] is a small programmable device with input and output pins and integrated or available addition of ADC controller and digital-to-analog (DAC) converters for analog outputs. They typically perform primary signal conversion and processing and execute built-in control sequences for local automation and alerting and have the ability to send data to other elements of the control hierarchy.

Some of the discrete sensors and actuators may perform internal signal conversions and processing. Such devices appear as addressable points for digital communication in the system hierarchy, usually as end or slave devices in a network segment controlled by a PLC master. Bus masters can participate in potentially multilevel configurations of control loop segments. Various forms of industrial buses and protocols – such as Fieldbus, Modbus, and PROFIBUS [6] – are used for communication with other elements.

Hierarchies of such control loops form what is referred to as distributed control systems (DCS) that may include a supervisory control and data acquisition (SCADA) systems for visualization and product line or plant-level control.

From the IoT point of view, existence of such systems provides the already established infrastructure for sensor and actuator systems that are digitized and may be interfaced via gateways. This may be performed at different levels in the hierarchy, including individual PLCs, clusters that form bigger control loops, or at the level of a production line or major building HVAC components using more elaborate protocols such as the BACnet [7].

To interface to legacy and DCS systems, many commercial IoT gateways include physical interfaces commonly used by industrial automation, such as the serial RS-485, RS-232, RS-422, and Industrial Ethernet. They are usually coupled with protocol converters for legacy industrial communication links and protocols such as Modbus and PROFIBUS.

Depending on where and how it is used, addition of IoT to existing automation may be deployed as a form of retrofit or enhancement of existing industrial automation installations. Primary benefits of adding IoT to the existing systems include the ability to collect process data and to aggregate them for a more holistic view and optimization at the factory, at the enterprise, and even at the cross-enterprise level. For example, Internet connectivity can provide machine operational data directly to the manufacturer and allow them to aggregate data over their installed base and use it to train and improve algorithms for fault detection and preventive maintenance. Additional insights provided by IoT technology can be used to improve operation of the plant by modifying control loops and set points of automation. In such a hybrid system, low-level control is usually left to the existing automation installation because of existing interfaces to devices and machines, and its autonomy and ability to continue autonomous operation even during periods when Internet connectivity are interrupted.

Interfacing Sensors in Maker Systems

A popular tool for experimenting and prototyping with sensors quickly and inexpensively are the single-board computers designed for the so-called maker community, such as Arduino [8] and Raspberry Pi [9]. They have a number of I/O pins for connecting digital and analog inputs and outputs, some of them configurable to work in either direction. They also include support for serial inputs and outputs, such as I2C and SPI buses. When provided, one or more analog inputs are connected to an ADC converter. Signal ranges usually correspond to the board power supply levels which are commonly 3.3 V or 5 V.

A great variety of affordable sensors is available for maker boards, such as the Seeed Studio [10]. They typically have signal ranges already adjusted to what the boards can handle and come with connectors mechanically matching the board I/O pins or their I/O extender "shields," thus allowing for direct connection and often circumventing the need for additional signal conditioning. Software libraries that acquire and convert inputs to engineering units are available for popular sensors and operating environments, such as the embedded Linux and Arduino integrated development environment (IDE). Preconfigured starter kits offer a combination of maker boards and matched sensors. Moreover, several IoT platform providers – such as Amazon Web Services and Microsoft Azure discussed in chapter "IoT Platforms" – provide edge software agents for some popular combinations of boards and sensors to capture data and to connect to their cloud offerings. This provides a quick and relatively easy way to get the real-world data to the Internet and to IoT back-end systems without having to deal with the details and intricacies of sensor data processing. This approach is suitable for learning and rapid construction of IoT proofs of concept.

Edge-Node Functionality

IoT edge node and thing implementations provide a wide spectrum of functionality, from the minimalistic data acquisition and transfer in integrated sensors to sophisticated data processing, storage, and analytics in high-end gateways and fog nodes. Most of them provide a similar sect of functions such as acquisition, processing, and transmission of sensor data. Those functions are described in detail in the remainder of this section. To simplify exposition, we describe them using an IoT gateway as the representative node and description vehicle as it can embody all of those functions and perform them on behalf of multiple endpoints in its care.

An IoT gateway links sensors and things at the edge with higher levels of system processing hierarchy and the cloud. It is commonly the point where Internet connectivity is achieved. A gateway is also a security boundary between things with varying levels of security and the secured IoT processing IT infrastructure. By virtue of its placement and function, an IoT gateway can also be an interface and a boundary between production- and process-level operational technology (OT), such as PLCs and control loops, and the IoT information technology (IT) part where most of the advanced data processing and storage takes place.

As its name implies, the basic and traditional function of an IoT gateway is to act as a communications bridge between sensors and actuators at one end and the Internet and cloud on the other. This entails existence of the appropriate hardware interfaces and processing logic for sensor data acquisition and actuation as well as protocol converter engines for "south" side wired or wireless communication links that it may support. On the "north" side, it means hardware interfaces for the types of uplinks that are supported, ranging from Wi-Fi and Ethernet to long-haul communications such as telco lines with the appropriate modem circuitry. The uplink communication is usually the Internet compatible at some point, so gateways commonly include the appropriate protocol stacks for TCP/IP and HTTP often with support for Transport Layer Security, such as TLS.

The primary function of an IoT edge node is to perform or assist data collection and to provide the connectivity and security for transfer of sensor data to other system components. As indicated earlier, data may be acquired directly from sensors and transducers or from things with digitized data sources via wired or wireless links that may or may not be IP-compatible.

Common data plane functions of a complete functional edge node, such as a gateway, may be categorized as:

- Core functions: data acquisition, transmission, and actuation
- Optional functions: data storage, event and alert processing, and control (automation)
- Advanced functions: analytics

In addition, the gateway performs control plane functions

- Security and system management

In the sections that follow, we describe the major data plane functions followed by an overview of the control plane functions.

Data Plane Functions

Sensor data are acquired at the edge by smart sensors and simpler sensors connected to gateways. Processing rules and policies in effect determine the relative significance and ultimate destination of the data. Consequently, some data samples may never leave the edge and be processed locally or discarded. Data and events that are destined for additional processing elsewhere in the IoT system may be communicated to peer nodes and to applications and services elsewhere in the system.

Major functions that may be supported by edge nodes such as gateways and fog nodes are depicted in Fig. 2.3. Gateways include provision for sensor interfaces, which fog nodes may or may not include, depending on their placement in system hierarchy. Boxes in the figure represent modular functional blocks that are configurable to form the pipelines necessary for the desired forms of processing of data at the edge. For instance, the simplest path of acquiring data for processing in the cloud could be formed by connecting the output of the data acquisition box to the input of the external communications module. A more involved path may add analyzing the data to detect event and alert conditions, storing copies of data locally, and running the data through the local control scripts that may result in actuation, possibly with the aid of the local analytics module.

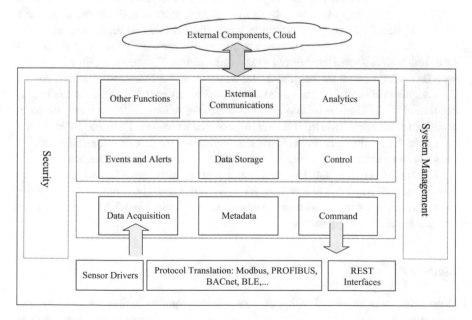

Fig. 2.3 Edge-node functional components

Data Acquisition

Data acquisition refers to sampling of sensor inputs at rates that may be fixed or selectable via configuration settings and changes. Time stamp of data capture may be recorded and associated with data items at this point. As discussed earlier, sensor data need to be linearized and converted to values useful for processing and exchanges with the rest of the system, such as common engineering units of measurement. That means that gateway data acquisition modules need to complete whatever sensor-specific processing steps may be required. In Fig. 2.3, a sensor driver function is in charge of processing data from raw and nonstandard sensors, and it may include sensor-specific driver modules for that purpose.

IoT gateway hardware based on widely available components from the PC world rarely support direct attachment of raw sensors. Implementations of IoT system that use them need to rely on smart sensors or use of intermediaries, such as PLCs, to provide transducer interfaces and most of the preliminary signal processing. Sensors and intermediaries can use a variety of connections and protocols to communicate data and commands.

Some of them are depicted in Fig. 2.3 labeled as the protocol translator box. An IoT gateway needs to support protocols used by the device types that are connected to it. Protocol translation means that the gateway needs to be able to parse and interpret received data packets and convert them to the appropriate common internal format used by the data acquisition module. On the way down from the gateway, protocol translators need to convert commands from the common gateway data information model to the format used by the target recipient, such as the PLC or an actuator. In addition to actuation, output commands may include controls and configuration, such as instructing a PLC to report its data or setting the sample rate for its sensors.

IoT data from individual sources are successive samples of sensor values, and thus, time-series data often referred to as data channels. Their capture time is marked by a time stamp either by the originating entity or by the gateway upon entering the data acquisition module. One of the design decisions in IoT systems is the implementation of the clock synchronization mechanism among different components in a manner that meets system timing resolution requirements. Data and their associated time stamps are then funneled to other functional modules configured to receive them, such as external communications, data storage, or events and alerts.

The functional box labeled REST is a generic representation of interface to things and gateways in the lower layers of system hierarchy. Such nodes may use their own data formats and functions and expose them via REST APIs over the compatible links and protocols.

Metadata

Sensor data may be tagged with metadata at various points in time between capture and transmission. Metadata refers to the management and storage of primarily contextual information that may be used to annotate data and be reported to external parties on demand. Simple forms of metadata include sensor types, engineering units in which they report data, range, accuracy, and minimum and maximum values. Other forms of metadata may include manufacturer, model and serial numbers, location of sensors, and the edge node. Implementation of this function usually requires a durable store for metadata entries with identifiers that can correlate them with the data sources or channels that they pertain to. A set of APIs is usually provided to query and, as appropriate, modify metadata.

The metadata definition can come into play at various stages in the system life cycle, such as manufacturing and installation times. Much of the metadata is static, so it does not need to be included in regular data reports and can be made available in response to specific requests. In current IoT practice, metadata are not well-defined or structured, and their treatment tends to be somewhat haphazard and vary across installations. Types and nature of IoT metadata are described in detail in chapter "IoT Data Models and Metadata", examples of its formal specification as used in standards proposals are outlined in chapter "IoT Data Standards and Industry Specifications", and an example of its use in a system is provided in chapter "Putting It All Together".

External Data Communication

The external communications module in Fig. 2.3 is in charge of input and output communications with the external parties, such as other peer nodes at the edge, fog, and cloud levels.

Inputs received by the gateway can include a variety of items, including data and state from peer nodes, commands for actuation, and policies and settings for various functions. The latter may include event policy updates, data sampling and retention settings, control script changes, and model or algorithm updates for analytics. External requests and APIs to gateway functionality may be parsed at this point or routed to the appropriate functional module unless they expose their own access points directly. Loading and updating of software modules and security updates are generally handled as a part of the control plane functionality described later, and they may use separate communication paths and credentials for that purpose.

The most common payload in an IoT system is reporting of sensor data. Unless the same information model or framework is used at both ends of the transfer, it translates the outgoing data into the format agreed upon with the receiving end. Message serialization (for outgoing) and deserialization (for incoming) are also commonly performed at this point. Serialization defines how individual bits and bytes appear in actual transfers over communication links, colloquially referred to as "the wire". A portion of such message may look something like this:

```
{"id":"150a3c6e-bef0ee0e", "temp":"n:77.6" "unit": "°F",
 "DateTime": "t:2019-07-15T14:50:07Z UTC"}
```

The message consists of "name":"value" pairs expressed in JSON-style notation both of which are commonly used in IoT systems. In this example, the reported value is "77.6". The associated metadata fields indicate that it is a temperature sensor reporting in units of °F. The time stamp indicates the date and time when this reading was acquired, using the UTC time zone. Details of IoT data formatting and metadata annotation are described in chapters "IoT Data Models and Metadata" and "IoT Data Standards and Industry Specifications".

Messages exchanged between IoT nodes may be delivered using push or pull modes. In the *pull mode,* a message is explicitly requested from the source (IoT server) by its recipient, an IoT client. This is typically accomplished via an API-styled "get data" for a specific data source or a group of related ones. In general, API exchanges require some synchrony between the sender and the receiver for the request to be handed out. The reply may be handled in a synchronous manner, with the requester thread pausing while waiting for the reply, or asynchronously where the recipient is notified when the reply is received.

In *push notifications*, the data source and the gateway acting on its behalf send data messages to an agreed point when the specified conditions are met, such as arrival of the data sampling time mark, or exceeding of the measurement threshold. Push notifications may be synchronous, such as a post via HTTP PUT to the receiver's specified port or address.

Asynchronous push notifications often take the form of *messaging*, where the sending process deposits the message into a specified shared queue, from which the receiving process retrieves it when ready. One of the major advantages of asynchronous message delivery is that it decouples timing of activities of senders and receivers. The data exchanged between the two parties are in a message, so their address spaces are also decoupled, and they may reside on the same or different machines. This is very convenient in dynamic distributed systems, such as the IoT, as functional modules of a data processing pipeline can be placed as needed and even migrated dynamically between machines in response to load variations without changes in coding or even disruption of the service when using messaging for data transfers.

A form of message exchange known as the *publish-subscribe* system, which was originally introduced on the Internet, has gained wide acceptance in IoT systems. In this approach, data sources "publish" data to a known place. Authorized clients can "subscribe" to data of interest and receive related messages at the specified rendezvous point. Publish-subscribe mechanisms have a number of advantages, including decoupling of senders and receivers, asynchronous communication, and scalability.

Publishers do not need to know the identity or number of their receivers (subscribers), and the receivers do not need to know the identity or state of their publishers. Moreover, a publish-subscribe mechanism allows 1 to 1 and 1 to many forms of distribution. The latter can be of great value to constrained nodes that may have multiple clients for their data but need to send their reports only once to the message queue where it gets distributed to multiple active recipients. Asynchronous communication means that message exchange does not require simultaneous presence of both parties, thus allowing constrained nodes to conserve power by entering sleep states after sending their reports.

In terms of implementation, publish-subscribe requires management of message queues in the infrastructure and servicing of the receiving and delivery ends. Implementation is often accomplished in the form of message brokers that manage lists of active subscribers and publishers in accordance with publish and subscribe requests. Brokers also provide queue management and storage that facilitates scalability and robustness, often with the addition of security and authentication of senders and receivers.

An edge node wishing to use one of the messaging or publish-subscribe systems needs to implement the client side of the corresponding protocol and use it to interact with the designated message broker that needs to be instantiated in the system infrastructure, typically at some higher-powered node or in the cloud.

There are a number of messaging and queuing system implementations that can be used in IoT systems. They include AMQP [11], DDS [12], and XMPP [13]. The most popular choice is the Message Queuing Telemetry Transport (MQTT) [14] which is described in more detail in chapter "Communications".

In addition to the core functions related to data acquisition and transfers, edge nodes may perform other optional functions, such as data filtering and event or alert generation, local storage, local control, and front-end analytics. They are described in some detail in sections that follow.

Events and Notifications

Events and notifications functions process incoming sensor data to detect conditions and state changes that require special handling. While there is no generally agreed-upon definition, events may be regarded as any observable occurrences in the system, while alerts are the more significant or critical events that may need immediate attention. For example, a temperature reading that exceeds a threshold may be treated as an event that needs to be sent as a notification to the cloud. However, a temperature exceeding maximum allowable range may be an alert that needs to be sent as a notification to the operator's console and/or an email or a text message. The classification of events and alerts is system and data-source specific. In a sense, all data readings are events, but only some of them need special treatment beyond the routine scheduled reporting to subscribers. State changes of binary nature, such as a valve being opened or closed, may be treated as events or simply preconfigured to always be reported to the cloud when detected. Alert notifications, when required, may be sent as notification messages from the gateway or forwarded for processing to a higher tier that aggregates groups and combinations of such messages and provides a notification service to the specified recipients.

Local Control

Local control capability is commonly provided in systems where latency is critical and/or edge autonomy is required. Low latency is ensured by the local execution of control due to its (network) proximity to data sources. Autonomy refers to the requirement for some control functionality to remain operational even during intervals when the node is disconnected from the rest of the system and the cloud in particular.

Control, when implemented locally, enables the edge node to execute pre-defined control sequences that generally cause local action when a specified condition occurs on some combination of local data. All input data that can potentially impact control need to be streamed to the control module. As each new input arrives, the control sequence is re-evaluated for potential action or sequence of actions. Control algorithms are usually defined as a set of rules in the form of scripts. PLCs use several control programming languages that often mimic construction of engineering diagrams, such as ladder logic. In consumer uses, such as home automation, gateways or ensembles of devices can engage in coordinated scripted group behaviors using tools such as If This Then That (IFTTT) [15]. Examples include the so-called scenes where lights and heating/cooling are adjusted accordingly when users are leaving or approaching their homes.

There is no standard language for control programming or scripting on IoT gateways in general. However, a flow-based visual programming tool, called Node-RED, has become quite popular. Node-RED is officially positioned as the flow-based programming for the Internet of Things. It was originally developed by IBM, and it is now available as an open-source project [16]. It provides a graphical user interface to construct data flows with the functional processing modules and actuation output stages. Node-RED framework then generates Node.js code that implements the corresponding actions and completes event and stream processing pipelines. The pipeline approach simplifies stream and event processing by providing modular processing functions and data connectors for external interfaces as well as structured interfaces between internal blocks for composition of more complex flows and functions.

A Node-RED system can combine multiple inputs and produce multiple outputs to and from its functional blocks. A number of common functions are provided in Node-RED libraries, and users may add their own custom modules as necessary. Connectors are available for some common types of IoT inputs and outputs, ranging from I/O pins on maker boards, industrial protocols like Modbus and OPC, and messaging via MQTT to human postings such as tweets from or to specified URLs. Node-RED has a community of contributors who provide a public library of thousands of open-source flows and nodes.

Figure 2.4 illustrates a simple processing pipeline constructed via Node-RED GUI [17]. In this example, it receives an event from a device, extracts temperature reading and classifies it, based on the reading and pre-defined ranges of tolerance, as safe or dangerous, and reports its status accordingly. In the extreme case of temperature, the script initiates a command to trigger the device shutdown, which may also include reporting of the corresponding status. Either the danger state or the

Fig. 2.4 Node-RED flow example [17]

shutdown trigger may be regarded as significant events to be highlighted on the system dashboard and possibly forwarded to an operator notification system such as twitter or email. Those actions may be provided by defining the additional functional blocks that are not shown in Fig. 2.4.

Data Storage

Data storage for the acquired sensor data samples may be provided by the gateway. Implementations tend to vary from simple circular per-channel buffers to sophisticated distributed time-series databases with varying resolution. An example of the latter is provided by the respawn system [18]. It provides a multi-resolution scheme where raw data is stored locally, and its up-samples, such as successive time-window averages, are computed and stored as coarser representation of data channels. Predefined levels of up-sampled data are sent to the cloud for caching and permanent storage. Query-dispatching modules keep track of the location of data and replicas at the various levels of granularity. This allows fast execution of range-based queries that can be satisfied from the nearest copy. Long-range queries may return large volumes of data efficiently supplied from the cloud. Queries requiring fine-grained resolution of specific data may be routed by the dispatcher to the lower-level leaf nodes, such as the gateways, where they reside. The up-sampling and cloud updating process is automatic, thus giving a wide flexibility of data access in terms of resolution, performance, and network bandwidth usage.

In general, the existence of local storage provides the flexibility to conserve network bandwidth and to reduce the load on cloud resources by selectively reporting data that meet certain conditions. This can be very useful for analog inputs that tend to drift so each reading may be different but not necessarily with sufficient magnitude to qualify as a reportable event. Depending on the nature of the physical thing being measured, only some of the data points may be of interest. For example, when monitoring a temperature in a building, successive minor changes may be of little interest, but those that exceed a comfort guard band or change by a significant percentage or amount from prior readings are. Qualifying readings may be defined as events and reported only when they occur, with raw readings stored in the local database should they be required for auditing or subsequent analysis.

Databases in edge nodes, when used, should be of the type suited for time-series data that can have a variety of different formats and time scales. Considerations involved in selecting types of databases for use in IoT systems are discussed in chapter "Cloud" in conjunction with cloud processing.

Edge Analytics

Edge-level analytics processing is becoming more common on the more powerful gateways and fog nodes. These can be systems that monitor key indicators to detect failures of the attached equipment, such as pumps or motors, or predict them for informed and timely preventive maintenance. Advantages of edge-level analytics include proximity to data sources, the ability to process high-frequency data samples with low latency, and conservation of network bandwidth by not sending them to the cloud. Moreover, edge analytics can run in real time and can operate even in the disconnected mode. For example, a self-driving vehicle executes its AI algorithms locally to navigate its path in the real world. It cannot depend on the cloud connectivity to do so, because any interruptions or prolonged latency could result in the lack of timely control and turn-by-turn guidance that could lead to accidents.

Since AI and ML algorithms need large amounts of data to be successfully trained, edge analytics is usually developed to work in conjunction with its counterpart in the cloud. As described in chapter "Cloud", ML and AI algorithms, such as neural nets, are developed in the cloud using its massive data aggregations and compute power. Portions of the developed algorithms, such as inference engines, can be placed at the edge nodes that can support their execution. The two systems tend to work in tandem, with the edge executing algorithms locally as in the self-driving vehicle example. While doing so, edge nodes in vehicles can record the scenes and situations that they have encountered and forward to the cloud the interesting ones, together with an indication of actions taken and their outcomes. More importantly, they can record situations in which the algorithm had difficulty or made mistakes so that the cloud can use them to improve the algorithm. The cloud portion of the AI system can collect data from all connected cars in use, several hundred thousand from one manufacturer alone [19] to improve training of its AI algorithm for self-driving. The improved algorithm can be distributed to all cars and updated using their Internet connection. This kind of cooperation can create a virtuous circle where more cars on the road and hours of recorded driving lead to better algorithms, which in turn results in cars that are more useful and valuable to their users.

Deployment of analytics is being accelerated by the new type of relatively inexpensive ASIC (application-specific integrated circuit) components that have been developed and optimized specifically for execution of AI algorithms at the edge nodes, such as [20] and others.

Function Placement in IoT Systems

Early implementations of IoT systems tend to have data collection at the edge, with most of the processing done in the cloud. This was partly due to the coinciding rise in cloud computing and the smart-phone design pattern of implementing services – such as voice assistants and activity monitors that provide mostly data collection and human interface on the user device at the edge with data processing done in the cloud. This approach has several drawbacks when applied to some IoT use cases, and the recent tendency is towards moving more functions to the edge. The right balance is system dependent and usually somewhere in between. As a general rule, IoT systems should be implemented using design tools and practices that facilitate flexible allocation of functions that is not fixed at the design time and allows late binding that can be finalized at the time of system installation and preferably even at runtime.

Some of the core edge functions, such as the sensor interfacing and data acquisition, may be anchored to the specific nodes, such as a gateway, by virtue of their physical placement and wiring. Other optional and advanced functions may be conceptually placed anywhere along the path from the data capture to delivery to its ultimate destination. In practice, this is determined by the system requirements and infrastructure capabilities as they are deployed and planned. In this section, we discuss the major considerations that come into play in making these decisions.

Optimal placement of data and processing functions is an age-old tradeoff in distributed systems. To a large extent, it is based on whether it is more cost-effective to move the data to the computation or to move the computation to the data. The answer tends to vary over time with the advances in technologies such as networking, computing, storage, and the resulting changes in the cost structure of components and services. Processing and storage used to be highly centralized in the era of mainframes, and then minicomputers and workstations started to move the pendulum towards distribution that accelerated with the pervasive use of personal computers. Cloud computing with its efficiency of scale and elastic resource allocation – fueled by advances in virtualization of computing, storage, and networking – led to cost-effective concentration of many functions back into the large data centers.

Major considerations in determining where and how to process data in IoT systems include:

- Availability and cost of bandwidth
- Latency requirements for time-critical operations
- Local autonomy of operations, including disconnected mode
- Security and data control or privacy concerns
- Cost and complexity of managing distributed nodes with computing and storage

When ample and cheap bandwidth is available, as in some Internet connections to the edge, it can be reasonable to just collect the data at the edge and do the processing in the cloud. Due to the scale and efficient usage of resources, storage and computation are generally considerably cheaper in the cloud. Due to high concentration, they are also easier to manage and allocate and to pay for on the as-needed basis. Additional advantages may be obtained when data are pooled in the cloud and made available for sophisticated and resource-intensive processing such as ML and AI.

On the other hand, when bandwidth is more limited and expensive, say over metered data links such as cellular, it makes sense to do at least some of the data processing and storage at the edge and to reduce the amount of data sent to remote locations. In some IoT uses, only a few data or significant events are of interest for further processing, but constant monitoring is required to detect them.

An example is provided by voice-activated assistants that have to permanently sample data from a microphone at kHz frequencies to detect utterance of the activation word or phrase. For that application, millions of samples that may be collected when the word is not uttered are of no use to the application. Sending them to the cloud would incur bandwidth costs and use of cloud resources with no tangible benefit.

Additional requirements in IoT systems that influence the function-placement tradeoff [21] include latency, local autonomy, and data control. Signals and events that have low latency requirements may not be possible to process in the cloud, which can have on the order of hundreds of milliseconds round-trip communications delays and rarely guarantee quality of service. Local autonomy, i.e., the ability to function in cases of loss of connectivity or cloud outages, is essential for many control applications, such as industrial automation, transportation, agriculture, mining, and power systems. This is generally achieved by having at least a minimal implementation of local control loops that can keep the system functional or gracefully degraded when disconnected.

Data control refers to some user requirements that some types of data must stay on their premises or within a specific geographic region. In addition to the obvious end-user considerations, this may be the case with manufacturing or building operations that want to safeguard the secrecy of their processes and efficiency of their operations (or lack thereof).

An important practical consideration in deciding on component and function placement in an IoT system is the comparative cost and complexity of installing and maintaining equipment at specific locations. Equipment in locations that are unattended or in harsh environments is generally more costly to maintain and harder to secure.

System functional modules are mapped to the infrastructure, and they work in concert to fulfill its overall design purpose. The infrastructure design determines node placement, power, and connectivity in accordance with the requirements defined by the problem that it is intended to solve. In practice, there are not that many options for the physical node placement other than (1) at the edge, near data capture, (2) on corporate premises, (3) in communications infrastructure, and (4) the cloud.

For example, in a manufacturing operation or a smart building, edge nodes can be placed on the premises and close to sensors if they use wired connections. The next aggregation level in the system hierarchy may be placed on corporate premises, which is a common preference due to security concerns and the ability to keep the data within a trusted physical boundary. The next level, when implemented, is typically the cloud where the physical equipment is usually rented in some fashion from a third party that owns and operates it.

One other potential option for placement of computational capability is within the communication infrastructure, such as in or in the vicinity of telco cellular towers when they are used for long-haul IoT links. Modern cellular communications usually require some distributed processing capability for call management and support of roaming users. That equipment is usually placed close to the cell towers since that is where the data sources and controls are. Cell tower sites tend to have fairly reliable electric power and are usually physically secured and monitored. This gives rise to potential placement of intermediate levels of IoT nodes, such as fog. Some companies are working on creating mini data center facilities with space or equipment to rent around cell towers. However, their business model and value are still being worked out so it is not clear if and when they may become a common option for placement of IoT nodes at intermediate levels of hierarchy such as fog.

As a good design practice, IoT systems should be enabled to allocate functions dynamically to the available infrastructure components anywhere between the edge and the cloud in response to system needs and requirements even as they may be changing in operation.

Fog Nodes

One other type of node sometimes encountered in IoT system descriptions is the fog node. The name is intended to imply that it "brings cloud to the ground". Fog nodes provide additional points in the functional continuum between the edge and the cloud by placing additional compute and storage capability closer to the edge. Functionally, they can perform what we referred to as the optional and advanced types of edge processing such as the more elaborate forms, local control, data reduction, filtering, and front-end analytics with potentially lower latency than the cloud. They are generally designed as more powerful nodes with virtualization and containerization capability that allows them to execute services that can be loaded as needed in response to processing requirements and load variations. At least in principle, fog nodes are also designed to support rebalancing of services and applications to meet the changing load and functional requirements.

Figure 2.5 illustrates a multi-node fog hierarchy as depicted in reference architecture from the OpenFog Consortium [22] that has since joined the Industrial Internet Consortium. It indicates that fog nodes are intended to be able to operate in multilevel hierarchies, which each higher level having progressively increasing levels of data aggregation and functionality. The top-level fog node may or may not reside in the cloud. In other words, fog hierarchies may be terminated on enterprise premises in situations where it is more cost-effective or there is a requirement for some critical data to not leave the company bounds.

Fog nodes tend to be more powerful but are otherwise not architecturally or functionally different from the edge gateways. They are part of the IoT functional continuum and can exist in one of the discrete points of IoT node placement discussed earlier. Use of fog nodes in the cloud refers to the functionality that matches and expands upon what is available at the lower tiers. However, communications with such nodes may experience higher latency due to their placement. Technically, edge gateways provided by cloud vendors as discussed in chapter "IoT Platforms" functionally resemble the configuration depicted in Fig. 2.5. From the design perspective, it is more useful to think in terms of which functions should be placed where and why, not so much in terms of what the nodes that host them may be called.

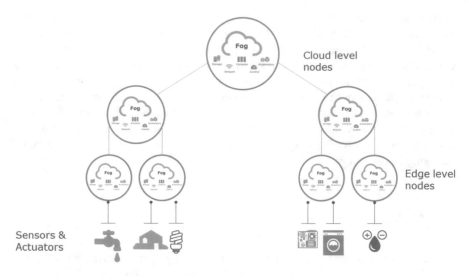

Fig. 2.5 Fog nodes in a multitiered hierarchy [22]

Control Plane: Security and Management

Gateway functions described up to this point are part of the data plane which performs the productive work of acquiring and processing sensor data and commands. They are active during the operating stage of a node lifecycle. As described, fully functional gateways may be fairly sophisticated computational and communications nodes that host and execute a variety of services. This means that they need to be properly secured and managed as any other piece of IT equipment of comparable complexity and importance.

Part of the IoT system that facilitates the initial bringing up of edge nodes and keeping them in secure operating states operates in parallel but separately from the data plane, and it is sometimes referred to as the control plane. Its primary constituent functions are security and manageability. They also operate in other stages of node life cycle, such as installation and provisioning.

The IoT management function is commonly implemented by complex systems residing in the cloud that can manage very large numbers of mostly unmanned edge nodes. Given the scale and generally little expectation of human assistance in the field, they rely on automation and policy-based management to perform their actions. In addition, IoT management systems may be tasked with additional group- and enterprise-level functions, such as inventory and asset tracking.

Management functions performed on installed and registered IoT nodes are described in chapter "Security and Management" and briefly overviewed in this section. In general, they are quite similar to the management operations in IT systems, and they tend to include:

- Fault management – troubleshooting, error logging, and recovery
- Remote monitoring, control, administration, and diagnostics

- Remote firmware and software updates
- Security updates
- Metering – network bandwidth and software usage

Since gateways have the processing power and network connectivity and are an endpoint of a management layer, they represent a logical and convenient point to provide additional management functions for their attached sensors and things. These could include status tracking, fault management, assistance, and caching for endpoint firmware and software updates. This function is not yet common in current implementations, but it should be incorporated as part of the IoT system control plane design.

Some IoT platforms include metering functions on edge nodes, such as network bandwidth and software usage. These may be used for several purposes, such as billing and throttling of bandwidth commensurate with available connectivity and cost. As an example, for mobile IoT nodes such as those mounted on trucks or containers, communications may be intentionally restricted when only costly connections are available, such as metered cellular. When the node finds itself in bandwidth-restricted zones, communication may be limited only to events and key data reports or just aggregates. When it is in the lower-cost bandwidth-rich zones, such as corporate-sponsored Wi-Fi, full data reports and software updates may be enabled.

Besides managing operational nodes in the system, an important function of the control plane is to facilitate and control the process of adding new nodes. This may happen in bulk during the initial system provisioning or incrementally when the system is in operation.

Adding a new node to an IoT system is accomplished through a process called provisioning that follows installation. Hardware installation includes both sensors and gateway devices. Sensors need to be installed in their place of operation and enabled for communication with the gateway. Depending on the system, this may entail direct wiring of sensor signals to a PLC or the gateway, connection of control buses used by the PLC, and establishment of wireless links appropriate for sensors that use them, such as ZigBee.

Initial bringing up of an edge node, including gateways and fog nodes, follows a more or less standard IT procedure of adding a new node – bootstrapping, connecting to the network, downloading, or locally applying BIOS and OS updates. The network connection may be a simple plugging of the cable into an Ethernet port or a bit more involved for link types such as a cellular connection that may entail installation of the SIM card and activation of the service. Depending on the system and on the available security enhancements such as Trusted Platform Module (TPM), some sort of trusted or secure boot sequence is generally deployed – as explained in chapter "Security and Management" in conjunction with security.

In order to join the rest of the IoT system, a node needs to be secured, authenticated, and registered. The process involves two parties and directions – the node being recognized as an authorized entity by other nodes and the cloud and the node itself authorizing external services to access it, issue commands, and download software for functional modules and OS/firmware updates. Node registration may involve a number of actions, such as entering into system and service directories or registries the node itself, its services and the data sources, and actuations for sensors and things connected to it. Node information supplied to the cloud may include its identifier, ways to address it on the network, protocols, and data formats that it supports. Node and sensor static metadata are normally furnished in this process, based on the information supplied during manufacturing and installation. In the reverse direction, the cloud may supply the new node with security credentials and designations of cloud access points for applicable message brokers and services.

A useful and often touted capability of commercial gateways is the so-called zero touch provisioning. It is a carryover from the networking gateway concept of being able to provision a gateway from the cloud following the physical installation, plugging in of network cables, and powering up. This capability reduces labor time and facilitates deployments with installers not proficient with IT skills. In principle, it also reduces provisioning errors – such as wrong data and labeling entries – by automating the process. It is especially useful in installations with large numbers of gateways. With IoT gateways, the provisioning process may be more complicated as it involves configuring and labeling their attached sensors and things. To assist with that part and the provisioning process, some gateway implementations provide a visual user interface with prompts and forms to assist installers in commissioning the system. It may provide automation combined with guided visual steps for configuring and registering sensors, adding metadata, and other provisioning functions. In general, this form of user interface should be able to operate between gateways and installer's devices using direct connections, such as USB, or short-range wireless connections such as Bluetooth for added security by virtue of requiring local presence and to eliminate dependency on fully functioning long-haul networking. Following installation, those links should be closed or restricted to reduce a potential security exposure.

Implementation of an IoT Edge Node

Smart IoT are generally designed for autonomous operation with the ability to interact directly with applications and other smart things in a peer-to-peer (P2P) or machine-to-machine (M2M) fashion. The idea is that things may discover each other and participate in ad hoc or preconfigured ensembles that cooperate in executing more complex group behaviors. For example, things such as devices and smart appliances may form groups, with new members added as they arrive, for sophisticated automation in homes. Much of the ongoing work on standards focuses on mechanisms for accomplishing this task with minimal or no a priori knowledge of the existence or capabilities of other nodes.

IoT edge nodes are basically computers augmented with sensors and actuators. Perceived from the outside, they may be viewed as a collection of defined network interface points and programming interfaces for accessing functions provided by the resident agents, such as obtaining sensor data and issuing actuating commands. These are commonly structured as APIs that other system components can invoke if authorized to do so. The totality of node functions at a given point in time is defined by its core services and whatever other modules and microservices may be active on it.

A machine-readable node description is supposed to indicate what the node can do and how to invoke it. Implementing those is important because node configuration and changes may be made at runtime, so its functions should be discoverable and not fixed at design time.

Some examples of generic functionality that an edge node may support and make available internally or externally via APIs on specified access points for the data plane include:

- Get sensor reading – single, group, all
- Get metadata – single endpoint, group, all
- Command – actuate output
- Operational settings and configuration, processing rules, subscribe/unsubscribe

The control plane may have different access points and include functions such as the following:

- Get software OS version, patch status, antivirus signature file date, etc.
- Download and install software update/patch.
- Update security credentials, add/revoke authentication, and authorization

IoT edge-node functions and interfaces are not standardized. However, IoT frameworks, as described in chapters "IoT Data Models and Metadata" and "IoT Data Standards and Industry Specifications", formalize specifications of some functions, their interfaces, and ways to describe and discover them among compliant nodes.

Implementation of an edge node requires assembling of the collection of functional modules to provide its required functions and the underlying hardware to provide physical interfaces and a platform to support their execution. In the sections that follow, we describe some of the aspects and considerations in IoT edge node hardware and software design.

Edge-Node Hardware

Figure 2.6 depicts a simplified version of key elements of IoT edge nodes. Those ingredients are present in autonomous "smart" nodes, gateways, and fog nodes, but with different level of complexity and capability.

Smart sensors and things may have integrated sensors and actuators that are part of an edge node. Higher-end function gateways typically do not have integrated sensors but may provide interfaces to them, either in the form of digital and analog I/O

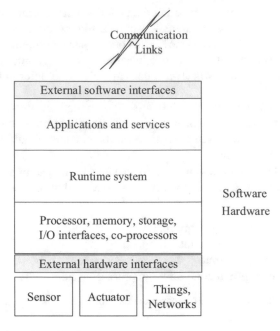

Fig. 2.6 Edge-node building blocks

described earlier in this chapter or as communications interfaces for buses and links used by devices and controllers that interface to sensors, such as RS-232, RS-485, RS-422, USB, Ethernet, and wireless.

Sensors may be connected using wireless links when wired infrastructure is not available or practical, to save cabling costs, or for user convenience. In uses such as agriculture and transportation, it is common to place battery-operated or solar-powered sensors in the field and link them to the Internet via communications gateways. Depending on the environment, range, and bandwidth requirements, wireless sensors can use a variety of links, including ZigBee, Thread, and 6LowPAN. For home and consumer uses, wide availability of Wi-Fi and BLE makes them a popular choice for many devices. Industrial applications may also use variants of hardened wireless communication where cabling is more expensive or impractical.

Edge nodes also include hardware interfaces for the "northbound" connectivity to other system components, the Internet, and the cloud. Depending on the intended placement and available networking infrastructure, these may range from local wired/wireless network links to wide-area networks, e.g., modems and SIM cards for variants of cellular networks, such as GPRS and LTE. Commercial gateways typically include wired and wireless Ethernet interfaces and some selections of medium- or long-range networking options packaged as variants of the basic model.

Simple and functionally specialized edge nodes may include integrated sensors and actuators, a microcontroller or a system-on-a-chip (SoC) with memory and I/O, and an external communications interface, wired or wireless. Higher-end edge nodes tend to be implemented using general-purpose CPUs with additional memory and storage capacity to support required system functions. As discussed later, their processors may be designed for extended operating range if intended for use in harsh environments. On the very high end, hardware components are chosen to assist virtualization and general-purpose operating systems with support for containers. Edge nodes may also include hardware accelerators for specific functions, such as FPGA (field-programmable gate arrays), or special purpose- built silicon such as ASICs for AI.

Edge-node designs may also incorporate hardware assists and specialized processor for security, such as Trusted Platform Module (TPM) and Trusted Execution Environment (TEE) that are discussed in detail in chapter "Security and Management" in conjunction with security.

Physically, IoT edge nodes may operate in a variety of settings, including harsh environments and remote unmanned installations. Nodes and gateways intended for use in harsh environments should use components with the extended operating temperature ranges, e.g., −40 °C to +85 °C (−40 °F to +185 °F). For industrial applications, they are often encased in hardened enclosures that can withstand vibration and protect the electronics from the exposure to dust and harmful elements in the environment. Enclosures may also be tamper-proofed to enhance security. Tamper proofing usually refers to a mechanism to generate an event or alert to the management console when the enclosure of the edge node is opened. Enclosures are commonly designed for fan-less operation and are usually rated and tested to some applicable industrial specification.

Edge-Node Software

Edge-node software needs to support connectivity, as well as command and control functions commensurate with its intended purpose. At the minimum, this implies having device drivers for the attached sensor(s) and communication links, supporting required protocol stacks, performing sensor data acquisition and signal processing, and managing actuation outputs if available. An edge node intended for direct connectivity to the Internet needs to support the IP stack and appropriate transport protocols, such as HTTP. It also needs to support an information model for representing data and metadata, coupled with the chosen serialization method for their transport, as described in chapters "Communications", "IoT Data Models and Metadata", and "IoT Data Standards and Industry Specifications".

Autonomous IoT are generally designed with the ability for self-describing and exposing public properties, access points, and methods. These are usually expressed using conventions of some standard or framework, as described in chapters "IoT Data Models and Metadata" and "IoT Data Standards and Industry Specifications". Clients and other things wishing to engage and interact with the thing must do so using the same framework or convention. Exchanges are preceded by thing discovery that may happen through network probing, announcement, or via device directories.

Runtime System

In general, a runtime system provides management and allocation of hardware resources, language libraries, and system services for executing applications. System services may include drivers for networking hardware, TCP/IP stack and Internet protocol implementation, and file system management for storage. A runtime system may also provide isolation to improve integrity and security of individual applications.

Implementations of edge-node software functionality vary in terms of complexity and flexibility. For limited-function things, all of their software may be implemented as firmware for the integrated microcontroller or SoC and stored in nonvolatile storage, such as flash. This approach tends to be cost-effective for high-volume products, but it can severely limit the flexibility in terms of adding or changing functionality later on. In more complex and higher-powered nodes, runtime services are usually provided by an operating system with the appropriate language runtimes.

Edge-Node Operating Systems

For different reasons, edge-node implementations use a variety of real-time operating systems (RTOS) and embedded or full-function versions of general-purpose operating systems such as Linux and Windows. The choice is often determined by considerations such as performance requirements, cost, and ease of development.

Key advantages of RTOSs include real-time capability, comparatively small footprint on the order of KBs of RAM and ROM, and generally low cost with many open-source versions available. RTOSs tend to use preemptive, priority-driven schedulers that facilitate implementation of deadline guarantees for execution of time-critical system functions. On the negative side, those systems tend to have limited choices of development tools, language support, and bindings and limited sets of drivers and protocol stacks that they support. They can be difficult to learn and program which, combined with significant fragmentation of the RTOS market, tends to limit the pool of system developers available for any particular system.

On the high end of the functionality spectrum, IoT gateways and fog nodes may be built using general-purpose operating systems that can provide runtime support for executing a variety of services and applications in a shared environment or sequestered with the aid of virtualization and containerization technologies. They tend to have driver support for a wide variety of hardware, rich choice of languages and libraries, networking and protocol stacks, file systems, and the like. If commercial products or their derivatives are used for implementing the control plane, they usually have client agents already implemented for the dominant general-purpose OSs. All of these may increase developer productivity by having many useful services to build upon, a wide choice of languages to use for different modules, and a large pool of familiar developers for the most popular ones. They can significantly increase system flexibility to change or add functionality by downloading and deploying modules without disrupting other node operations. With support for containers, complex functions such as edge analytics may be dynamically loaded, activated, and even load-balanced across a group of nodes to meet the varying load and application needs. Containers and virtualization support runtime deployment of functions and can provide additional isolation for improved system security, resource allocation, and integrity in terms of containing damage from failures of individual components, containers, or partitions. For example, data plane and control plane, i.e., the security and management functions, may be executed in separate containers or virtual machines to increase separation in terms of performance and fault and intrusion containment.

On the negative side, full-function general-purpose OSs tend to require more powerful processors and significant amounts of memory and storage to run, thus making edge nodes that use them considerably more expensive. In addition, they tend to use scheduling algorithms that generally do not support execution-deadline guarantees for time-critical tasks which may be an issue in some applications.

Embedded versions of popular general-purpose operating systems, such as Windows Core and variants of Linux including Android, aim to reduce the footprint and improve performance by custom-tailoring distributions for smaller devices and eliminating system functions and services that are not needed. They tend to use less memory than their general-distribution cousins, but still typically orders of magnitude more than RTOSs.

Always-on and Unattended Operation

In many instances, IoT edge nodes are supposed to perform their functions continuously and in effect be always on. Their software needs to be designed accordingly and fulfil those requirements even for nodes that operate in unattended and possibly hard-to-reach locations.

Unattended operation means that the node should be able to assume full operational state upon being powered up, without any other intervention or dialogue requirements. This is a common practice in embedded systems and fixed-function appliances, but not in general-purpose computing that higher-end edge nodes may be based on. It also means that an edge node should reboot itself and resume a known good operational state after intermittent failures. Reboots can happen after loss of power and software or hardware errors. Resuming operational state means not only rebooting the operating system, if present but also activating all key services either from the last known good operational state or as specified in the default configuration.

Watchdog Timer

Fault detection and recovery may be aided by implementing a watchdog timer, usually as a combination of hardware and software. The *watchdog timer* (WDT) is a device implementing a recovery technique with origins in embedded systems. Its purpose is to force a reboot when the system encounters errors, typically software, that cause it to "hang," i.e., get stuck in an application or a system routine that renders it unresponsive and unable to run other tasks. Implementing a watchdog timer requires a special hardware provision and system software assistance. It basically consists of a hardware timer, often implemented as a single-shot trigger circuit and tied to the system reset power line or signal. There are variations in implementations [23], but the circuit basically operates by depending on receiving a succession of input signals whose inter-arrival times are below the specified timeout value. Those signals are produced in response to software activating the watchdog input pulse in appropriate intervals. Failure to receive an input pulse for the duration that exceeds the chosen timeout causes the watchdog timer to trigger and activate the system reset line. In essence, the timer acts like a form of a dead man's switch wired to trigger a reboot unless it is repeatedly told not to do so by a software-generated input signal. Many microcontrollers designed for embedded systems include WDT circuitry that operates independently from the processor. General-purpose CPUs, such as those found in desktops, usually do not include a WDT. To implement watchdog functionality, edge nodes and gateways that use more powerful processors need to include external WDT circuitry in their designs.

In terms of software, a simple implementation is to create a system task running at the lowest system priority when nothing else is ready or available to run, just above the idle state. With the properly designed timeout period, the task activating the watchdog pulse will run often enough to avoid triggering. In case of a system failure resulting in a hung state, the WDT reset task will not be able to run within the WDT timeout period, and the system will be restarted and restored to an operational state.

More elaborate implementations are possible with multistage WDTs that generate several timed signals in sequence before restarting the system. Those signals are a succession of imminent restart notifications that may be used to activate routines that record the detailed system state for post-recovery analysis and bring the external peripherals and machinery to a safe physical state before restarting the edge node. Such warning WDT signals may be connected to non-maskable interrupts to increase their chance of running when the system is operating in a degraded mode. The final signal then triggers the power reset cycle.

Applications and Services

Edge node purpose and use are primarily determined by the functionality provided by its data plane, such as sensor data acquisition and transmission, local control, and data storage. A good design practice is to implement those functions in modular fashion. This approach allows functions to be modified, added, removed, or relocated as needed. With proper runtime support, individual functional modules may be coded in different languages and deployed while the node is running. If each module provides its own interfaces, data processing and command pipelines can be set up and reconfigured to make use of additional modules and to bypass the removed ones.

The implementation of a data plane functionality results in a cooperating set of routines that closely interact and share a common purpose. They may collectively be viewed as a data plane agent. A separate agent or agents need to be provided to implement the control plane functions. This separation makes sense since agents operate independently of each other and generally exchange their data and commands with different entities in the cloud. In practice, security and management services will usually interact with a separate aggregation and control point with its own management console and operators that focus exclusively on matters such as security status and management of software updates. In the current practice, security and management systems are often built on versions of corresponding commercial IT products modified for IoT. They operate by installing their own agents on client (edge) platforms that interact with dedicated security and management servers that manage large groups of IoT nodes. While they may operate in a sequestered isolated and perhaps more trusted portion of the environment, control plane modules still need access to system services and applicable language libraries provided by the underlying runtime system.

Thus, higher-function edge nodes may contain an elaborate multi-module data plane agent and another one or two dedicated agents for security and management functions. They would fit in the side boxes depicted in Fig. 2.3 as security and management system.

Microservices

Microservices are a commonly used design technique in modular implementations of edge-node functionality. Microservices are a distributed system version of modular programming with roots in system-oriented architecture (SOA). They implement and encapsulate a particular system function and are relatively small and easy to understand and develop. Groups of microservices can interact at runtime to implement more complex services that can span multiple nodes and are scalable since individual components may be replicated to handle additional load.

Microservices are a popular design approach in implementing complex distributed systems that can be operated in cloud environments. In a way, they are a cloud incarnation of the modular programming style that emerged as an antidote to monolithic code decades ago [24]. The key idea is to break up the implementation of a larger system into modules that correspond to functional tasks and to hide their internal implementation details and information. Module functions are made available to other parts of the system through its interfaces. Obviously, this makes even more sense in distributed systems where different services may reside on different nodes that do not share address spaces. One of the key benefits of microservices and modular programming is that the overall system is viewed as a collection of interlocking functional blocks with hidden inner workings that should be easier to develop and understand than a large monolithic piece of code. In this approach, functional changes and modifications can be limited to one or a few affected modules, and they can be designed and maintained by smaller teams. Since microservices communicate only via APIs that can be serialized over common protocols, such as HTTP, they may be implemented in different languages to suit the needs and preferences of individual teams. In the current practice, microservices tend to be independently deployable in an automated fashion. They are an integral part of the so-called cloud native programming.

Such flexibility may come at the additional cost of implementing and managing many interfaces, registering and tracking modules, and the runtime overhead of multitasking and cascading inter-module message queues and interface entry/exit points.

An alternative is to implement edge-node functionality as a single monolithic block of code. This is quite commonly done for expediency when constructing proofs of concepts and pilots with relatively few sensors and basic data transfer functionality. When the system grows, there is a natural tendency to deal with the added scale and functionality by modifying the existing code. This temptation should be resisted, as the advantages of the modular approach over monolithic code far exceed its disadvantages when the system is scaled up. Thus, a major refactoring or a complete replacement should be accounted for in the scaling plans of such projects from the beginning.

An Edge-Node Implementation Example

As an example of the functions in an implementation of an IoT edge node, we describe a fairly comprehensive specification and open-source design project from the EdgeX Foundry [25]. It is intended to be hardware agnostic and to allow functionality to be distributed across multiple edge hardware nodes or across processors within a single node. It aims for cross-platform interoperability and supports multiple hardware architectures, operating systems, and application and programming environments. The stated project sweet spot are edge nodes such as embedded PCs, hubs, gateways, routers, and on-premises servers.

Figure 2.7 depicts the software structure of EdgeX IoT edge node. The figure illustrates key application functional blocks. The underlying software base, such as the operating system and runtime service environment, is not shown.

The "south" side of the node consists of interfaces and protocol translators for commonly used industrial buses, such as Modbus and BACnet, wireless links and protocols including ZigBee and BLE, and MQTT and REST as a generic interface to other gateways, edge aggregators, and smart things. SNMP is a very common management interface for IP networks and devices.

Functional modules depicted in Fig. 2.7 are implemented as microservices for modularity and composition.

Node services are layered and grouped into five categories: (1) core services, (2) supporting services, (3) export services, (4) device services, and (5) system services.

Core services, which are mandatory in all EdgeX implementations, include:

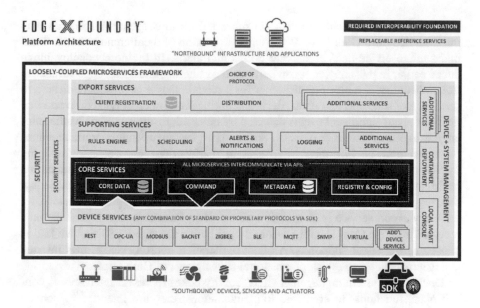

Fig. 2.7 EdgeX node functional components [26]

- Core Data – a centralized persistence facility for data readings collected by devices and sensors that other microservices can call upon. Sensitive, time critical, and data that do not need to be persisted can be streamed through it.
- Command – sends commands and actions to device and sensors on behalf of other local microservices, other local apps, and any external system, e.g., cloud. It exposes commands in a common, normalized way to simplify communications with devices. GET command is used to obtain data or status from sensors and PUT to take action or modify settings.
- Metadata – has knowledge about devices and sensors and how to communicate with them. It manages information about devices and sensors. It knows type, organization, and data reported by devices and sensors and knows how to command devices and sensors.
- Registry and configuration – the centralized management of configuration and operating parameters for all microservices and location and status of microservices.

Core Data uses an object-styled data model with events that contain device identifiers and one or more readings structured as name-value pairs. Readings are associated with value descriptors that essentially contain metadata such as data type, minimum and maximum allowable values, and labels. Data model is a rendering of a general IoT information model described in chapter "IoT Data Models and Metadata".

Optional supporting services include rules engine which implements rule-based edge event triggering mechanism to invoke local actions when predefined conditions are met. It can be attached to data via subscriptions to outgoing distribution data microservice or to Core Data. It can cause activation of local commands via command microservice and/or generate events and alerts. In the nomenclature introduced earlier in this chapter, it can perform data filtering and local, edge-level control. Scheduling is a bit of a misnomer for a module that is at present dedicated to scrubbing of stale and sent data. Alerts and notifications microservice delivers notifications about significant events detected by other microservices, such as the rules engine, to other systems or persons. For example, system operators or maintenance personnel may be informed when system malfunctions are detected. Logging microservice in this implementation deals primarily with system management and health events including traces and errors of the more traditional IT nature. Sensor data logging, if required, may be implemented using Core Data persistent store.

A key export service is distribution. It can filter, transform, and format the data received from the Core Data for external distribution through REST and MQTT as chosen by the client endpoints. A client registration microservice allows internal and external clients to register (subscribe) as recipients of Core Data.

Device and system management and security boxes represent interfaces to control plane functions. In complete implementations, they are intended to contain local agents and microservices for security and management functions.

References

1. Sampling Theorem [Online] Available at: https://en.wikipedia.org/wiki/Nyquist%E2%80%93Shannon_sampling_theorem (Accessed Dec 15, 2019)
2. Charles, P. (2016) *Encyclopedia of Electronic Components, vol. 3: Sensors*. San Francisco, CA: Maker Media.
3. Microsquirt Thermistor Quick Start Guide [Online] Available at http://www.useasydocs.com/theory/ntc.htm (Accessed Dec 15, 2019)
4. Siemens SIMATIC Wiring [Online] Available at https://www.automation.siemens.com/sales-material-as/interactive-manuals/getting-started_simatic-s7-1500/documents/EN/wire_en.pdf (Accessed Dec 15, 2019)
5. Bolton, W. (2015) *Programmable logic controllers*. 6th edn. Waltham, MA. Elsevier Newnes
6. Sen, S. K. (2014) *Fieldbus and networking in process automation*. Boca Raton, FL: CRC Press.
7. BACnet Standard, ASRAE 135-2016 [Online] Available at https://www.bacnetinternational.org/page/BACnetStandard (Accessed Dec 15, 2019)
8. Arduino [Online] Available at https://www.arduino.cc/ (Accessed Dec 15, 2019)
9. Raspberry Pi [Online] Available at https://www.raspberrypi.org/ (Accessed Dec 15, 2019)
10. Seeed Studio Maker Sensors [Online] Available at https://www.seeedstudio.com/ (Accessed Dec 15, 2019)
11. AMQP [Online] Available at: https://www.amqp.org/ (Accessed Dec 15, 2019)
12. DDS [Online] Available at: https://www.omg.org/spec/DDS/1.4/PDF (Accessed Dec 15, 2019)
13. XMPP [Online] Available at: https://xmpp.org/ (Accessed Dec 15, 2019)
14. MQTT [Online] Available at: http://mqtt.org/ (Accessed Dec 15, 2019)
15. If This Then That, IFTTT [Online] Available at https://ifttt.com/ (Accessed Dec 15, 2019)
16. Node-RED [Online] Available at https://nodered.org/ (Accessed Dec 15, 2019)

17. Node-RED flow diagram [Online] Available at: https://developer.ibm.com/recipes/tutorials/getting-started-with-watson-iot-platform-using-node-red/ (Accessed Dec 15, 2019)
18. Buevich M et al (2013) 'Respawn: a distributed multi-resolution time-series datastore', *IEEE 34th Real-Time Systems Symposium*. Vancouver, BC, pp. 288–297.
19. Tesla autopilot [Online] Available at https://www.tesla.com/autopilot (Accessed Dec 15, 2019)
20. Edge Tensorflow Processing Unit [Online] Available at https://cloud.google.com/edge-tpu/ (Accessed Dec 15, 2019)
21. Satyanarayanan, M. (2017) 'The emergence of edge computing' *IEEE Computer*, 50(1), p 30–39.
22. Open Fog Reference Architecture [Online] Available at https://www.iiconsortium.org/pdf/OpenFog_Reference_Architecture_2_09_17.pdf (Accessed Dec 15, 2019)
23. Ganssle, J. (2012) 'A designer's guide to watchdog timers' [Online] Available at: https://www.digikey.com/en/articles/techzone/2012/may/a-designers-guide-to-watchdog-timers (Accessed Dec 15, 2019)
24. Parnas, D. L. (1972) 'On the criteria to be used in decomposing systems into modules', *Communications of the ACM,* 15(12), 1053–1058.
25. EdgeX Foundry [Online] Available at: https://www.edgexfoundry.org/ (Accessed Dec 15, 2019)
26. EdgeX Foundry Wiki [Online] Available at https://wiki.edgexfoundry.org/ (Accessed Dec 15, 2019)

Chapter 3
Communications

Networking has two important parts of primary interest to IoT system designers: (1) the edge network that provides wired or wireless connections to sensors and things and (2) the Internet that interconnects all other parts of the system. Edge networks operate in a variety of environments and conditions that can pose additional challenges and result in somewhat unique implementations. However, data from the edge are ultimately destined for the Internet, and thus the IoT Edge network should be designed with that transition in mind. Internet is the network of networks based on the sound design principles that have withstood the test of time. Obviously, it makes sense to apply and emulate those ideas in IoT Edge network design.

In this chapter we describe layered network models and design principles that are the foundation of the Internet and most IoT networks. This is followed by the discussion of low-power lossy wireless edge networks (LLNs), long-range low-power networks, and telco networks in the commercial licensed spectrum as they pertain to IoT. The chapter closes with a presentation of techniques and protocols for integrating sensor LLNs into the Internet, including 6 LoWPAN and CoAP, followed by a discussion of IoT messaging and queuing delivery mechanisms and MQTT.

Layered Network Design

Layering is one of the fundamental principles of contemporary network design. It is the architectural foundation of the Internet with proven durability and scalability with billions of active endpoints. Its global infrastructure consists of numerous routers, switches and support services built by a multitude of vendors on very diverse software platforms and interoperating using the common public specifications.

Layering refers to the idea of implementing network functionality in the form of functional layers, each with a well-defined set of functions and interfaces to the layers below and above. Each layer uses the functions of the layer below it and

© Springer Nature Switzerland AG 2020
M. Milenkovic, *Internet of Things: Concepts and System Design*,
https://doi.org/10.1007/978-3-030-41346-0_3

provides services to the layer above it. Layers are independent thus leading to a simplified modular design. A standardized definition of layers allows independent implementations and ensures interoperability. This is particularly useful in Internetworking where different physical networks and protocols may be integrated at the networking layer to make their differences imperceptible at higher layers.

The basic elements of a layered model are services, protocols and interfaces. A *service* is a set of actions that a layer provides to a higher layer on the same machine. *Interfaces*, also called service access points, are the access points for services. *Protocol* is a set of rules that a layer uses to exchange information with its counterpart on the remote peer node that it is interacting with.

Figure 3.1 illustrates network layers that would typically be involved in message exchanges between two endpoints on the Internet, designated as hosts X and Y. Since it is a global network of networks, many intervening routers may be traversed by a message on its journey from the source to the destination node. Two of them with their corresponding network layers are depicted in Fig. 3.1 to illustrate key concepts.

The lower portion of Fig. 3.1 illustrates network layers involved in the process. Going from top to bottom, the layers are application, transport, network, and data link/physical. Layers are designed to interact with the corresponding layers on the two communicating nodes using mutually agreed upon protocols. For example, an application on one node will interact with an application on another node by using the application protocol.

This is a logical view of the system. In practice, each message traverses the entire stack of layers. That is, an application message will be handed down to the transport layer on its originating node and then on down until it is actually transmitted via the

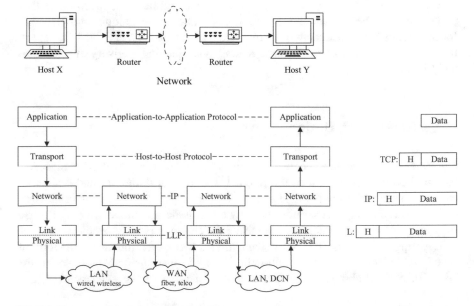

Fig. 3.1 Internet network layers and data flows

lowest layer that manages the actual physical communication link. As illustrated in Fig. 3.1, that is usually the nearest network neighbor on the packet's route to its destination node.

Individual layers use headers and sometimes trailers to provide addressing and other control information for layers on peer nodes to be able to process the corresponding protocol data unit correctly. Protocol data unit (PDU), shown in Table 3.1, is the term used to describe collections of bits that are actually exchanged between the two communicating layers. They contain the payload and the specific headers and trailers that may be used for packet identification, routing, integrity checking, and security. Figure 3.1 illustrates use of headers using the Internet Protocols TCP and IP, described later in this section. Trailers are omitted for simplicity.

In Fig. 3.1, the original message submitted for transmission by the Host X to its application network layer for delivery to an application at the Host Y is labeled as data. The application layer hands it down to the transport layer, using TCP in this example, that adds its own header. That whole expanded collection of bits, consisting of TCP header and the original data payload, is just the payload for the network layer below. Before sending it on, the networking layer appends its own processing header, IP. The same process, called *encapsulation*, is repeated at the lower layer until the bits are eventually transmitted on the link. At the receiving node, the process is reversed. Each layer processes its respective headers and trailers and strips them off before submitting what it regards as the payload to the next higher layer. The process continues until the original data payload is delivered to the application at the destination node, depicted as Host Y in Fig. 3.1. Figure 3.1 also illustrates the fact that some intermediate nodes, such as routers, may implement only a subset of layers, usually up to and including the network layer. Arrows indicating message flows in Fig. 3.1 are shown as unidirectional to illustrate the traversal of a particular message sent from the host X to Y. In practice, flows between layers and nodes are bidirectional, and messages can flow in either direction.

Table 3.1 Network layers and functions

Internet model	OSI model	IIC IoT model	Functions	Protocol data unit (PDU)
Application	Application	Framework	Data exchange between applications at endpoints, interoperable data	Data, information
	Presentation			
	Session			
Transport	Transport	Transport	Data segments between endpoints, segmentation, acknowledgments, multiplexing	Message (datagram), segment
Internet (network)	Network	Network	Packets transferred between endpoints, may not be on the same network, routed as needed	Packet
Link	Data link	Link	Transmission of digital frames between two endpoints connected by the physical layer	Frame
	Physical	Physical	Transmission of raw bit streams between two endpoints over a shared physical medium	Symbol

In addition to transferring data packets, various layers and protocols may perform additional control and management functions including:

- Packet integrity (checking?) management
- Congestion and flow control
- Routing
- Packet fragmentation and reassembly
- Packet reordering

Packet integrity refers to making sure that packets are delivered free of errors. This is usually accomplished by means of adding error detection codes. The sender computes some form of checksum on headers and data and prepends it or appends it to the outgoing packet. The receiver computes the checksum on the received data and compares it to the received checksum. Errors are detected when there are discrepancies between the two. Positive or negative acknowledgments from the receiver may be used to indicate the outcome, and the transmission may be repeated if reliability is required.

Congestion and flow control are the mechanisms used to throttle the sender's outgoing traffic, if necessary, to not overwhelm the receiver. This implies existence of some signaling mechanism between senders and receivers to adjust the data rate to what the receiver can handle. Congestion control usually refers to direct links, whereas flow control may be deployed on an end-to-end basis.

Routing is the mechanism used to determine where to send a packet if it is not addressed to any of the nodes on the direct outgoing links. On the Internet, routing is performed by the network layer. Network routers maintain routing tables based on connectivity and the routing metrics.

Packet fragmentation may be necessary if the submitted payload is too large to fit on a particular link due to packet-size limitations on the link itself or in the receiving node.

Packet reordering may be necessary in networks, such as the Internet, where packets carrying fragments of a message may be received out of order. This can happen since packets that carry fragments or segments of a message are sent independently of each other and may traverse different network routes with different delays on their way to the destination. Reordering of packets requires protocol support for marking the sequencing of packets by the sender which the receiver can use to reconstruct the original message in its proper order before delivering it.

Network Reference Models

Table 3.1 lists three reference models and their layers, functions, and protocol data units which are the types of messages that individual network layers exchange. The three presented reference models are the Internet model [1], the OSI (Open Systems Interconnect) model [2], and the Internet Industrial Consortium (IIC) IoT connectivity model [3].

The Internet model is the blueprint for IoT networking and the Internet itself. The OSI model was proposed in the early 1980s as the foundation for creating standards and implementations for networking and Internetworking systems. It was quite influential for a while as a model and an effort backed by standards bodies and governments, mostly in Europe. For a variety of technical and political reasons, one of them being that the specifications preceded implementations, OSI was not widely implemented, and it was not a commercial success. With the benefit of hindsight, we know that the Internet won, but that was by no means clear at the time. The Internet Industrial Consortium (IIC) IoT model is aimed at the IoT systems in general and the industrial IoT systems in particular, as that is the primary focus of the IIC consortium.

The primary observation is that, at this level of abstraction, all three reference models appear to be structurally similar, with nearly identical lower layers and different levels of detail in the specification of the application (top) layer. We discuss those in the subsequent section on the application layer.

The other noticeable difference is in the link layer. Internet definitions specify a single link layer and do not subdivide it further. In the OSI model, in much of the literature, and in many practical implementations, it is treated as two layers – link and physical. The IEEE 800 series standards that deal with network physicals, some of which are discussed later in this chapter, define separate physical and link layers. This distinction is mostly definitional as much of the Internet runs on networks and links with the well-defined physical and link layers. Ultimately, all the needed network functions have to be implemented somewhere in the system hierarchy, and the differences in definitions are really about how they are categorized and conceptually grouped together.

The Link Layer: Physical and Link

As indicated, the physical layer is in charge of transmitting raw bit streams on the wire between the two endpoints. These are the two adjacent nodes with a direct link, not necessarily the ultimate endpoints in the data exchange. This layer deals with physical issues, such as connectors and number of wires, and with bit signaling on the line, such as clock/bit synchronization, encoding of ones, zeros, and idle states. For radio networks, the physical interface deals with selection of frequencies and signal strength.

The link layer superimposes a frame structure on the raw bits on the wire. Frames usually contain source and destination addresses of nodes exchanging the frame. For Ethernet and many other networks, these are unique hardware identifiers such as Media Access Control (MAC) addresses. Each network interface, such as Ethernet, Wi-Fi, and Bluetooth, has a unique MAC address assigned by the manufacturer. For example, a PC or a smart phone may have several MAC addresses, one for each type of the network that they can connect to.

The link layer receives network packets from the network layer and adds some preambles and optionally trailers to form link frames that are transmitted over the wire by the physical layer. These transmissions are point to point between nodes on the same network segment or link with no routing.

On shared links, link layer may be in charge of obtaining access to the shared medium for transmission. On bus and broadcast radio networks, that may entail execution of an arbitration algorithm to coordinate and manage access when multiple nodes wish to transmit at the same time.

The Network Layer

The next layer in the hierarchy is the network layer. This is the Internetworking layer that can shuttle packets across intermediaries and different types of networks on their way from source to destination. The Internet was originally designed as the network of networks, and this integration happens at the networking layer. While both Internet and OSI contain a networking layer at this level, they differ in some aspects of their functional allocation. Being focused on IoT, we'll describe mostly the Internet version where the networking layer is often referred to as the IP layer.

At the network layer, each endpoint has a globally unique address, thus allowing for network-wide identification of source and destination nodes. On the Internet, these are known as the Internet Protocol or IP addresses. The original 32-bit IP address specification is usually referred to as the IPv4 for version 4. The newer IPv6 has 128-bit addresses as discussed briefly later in the section on 6 LoWPAN.

As Fig. 3.1 illustrates, network packets may traverse multiple segments on their way from source to destination. Links on the way may be heterogeneous, such as any combination or multiples of wired and wireless LANs, point-to-point cellular links, or wide-area networks (WANs). The beauty of the Internet and its layered architecture is that this is completely transparent to the upper end-to-end layers, such as transport.

Some of the key functions of the network layer include:

- Addressing
- Segmentation
- Routing

IP networks use IP addresses to uniquely identify endpoints of a packet exchange. Larger payloads may be split into smaller packets for subsequent reassembly before delivery. Packets are sent towards the destination independently of each other, and, depending on the link load and connectivity, they may traverse different routes. Routers and switches on the Internet maintain routing tables that, based on the connectivity and routing metric, indicate the next hop to which the packet is to be sent until it reaches the gateway or the LAN connected to the destination node. Routing tables are updated to reflect the changing network load, additions of new links and routers, and departures of those that fail or are decommissioned. Routing is done on a hop-by-hop basis, and the router knows on which link to send the packet towards the destination, but it does not know or try to create the complete path from source to destination.

IP strives for simplicity and flexibility, and it does not guarantee delivery of packets. They are sent on their respective routes on the "best effort" basis, but not otherwise acknowledged or retried if lost. A consequence of this approach is that packets containing parts of the same payload may arrive out of order or be lost due to errors or congestion along the way. Integrity of transmissions is the responsibility of the next layer up, the transport layer.

The Transport Layer

As Fig. 3.1 illustrates, transport is the first network layer that operates between the source and the destination of the payload exchange. Layers below operate on a hop or link basis. Internet has an end-to-end design philosophy that states that the endpoints should provide for integrity of transmissions. That means that they can provide reliable delivery when desired. On the Internet, two protocols are widely used at the transport layer – Transport Control Protocol (TCP) and UDP (User Datagram Protocol).

TCP is a connection-oriented, reliable, byte-stream service. It ensures that all data bytes sent are received error-free and in the correct order. Connection oriented means that the two endpoints need to establish a connection before the data exchange can occur. TCP manages segmentation and performs flow control if needed so as not to overwhelm the receiver. Maximum segment size is agreed upon between the two parties in the process of connection establishment. The TCP layer running at the receiving end checks received data packets for sequencing and errors and then discards duplicates and requests retransmission of the ones that are lost or corrupted. Received data are reordered as necessary to replicate the original sending sequence.

Another service that may be provided by the transport layer is multiplexing of traffic between sending and receiving ends. On the Internet, this is accomplished by means of ports which TCP and UDP use to identify sending and receiving applications. This sub-addressing mechanism in effect allows the establishment of communication subchannels for data exchanges between specific applications. TCP and UDP use 16-bit port numbers in their headers. Widely used Internet applications and services use and listen to the well-defined ports, such as the port 80 for HTTP traffic, 20 or 21 for FTP. On the Internet, ports 0–1023 are registered and reserved as the well-defined ports. Other port numbers are registered for temporary or permanent uses, and port numbers above 49,151 can be dynamically established and used by the specific applications in private arrangements.

TCP is optimized for accurate delivery, and it is used by applications that require it, such as World Wide Web (HTTP), email (SMTP, POP), and file transfers (FTP). Due to the need to retransmit or wait for the missing data, TCP may incur relatively long delays, on the order of seconds, that may be unacceptable for some IoT uses.

UDP, the other Internet Protocol at the transport layer, is an unreliable connectionless protocol. Somewhat akin to the regular postal mail delivery, it is unreliable in the sense that its datagrams (data packets) are sent and the best effort is made to deliver them, but receipt is not acknowledged. UDP does not provide delivery guarantees or duplicate protection. It does provide integrity verification in the form of checksums that allows the receiving end to detect errors. This mode of communication makes UDP fast and suitable for real-time uses and applications that can afford some packet loss or do not benefit from retransmissions. For example, there may be no value in retrieving and delivering a lost audio packet in a voice over IP application if the conversation has already progressed beyond its time frame. On the other hand, some video streaming protocols over the Internet use TCP at the transport layer for its delivery guarantees. They cope with delays by prefetching and buffering data before starting the playback to hide the latencies and to provide a smooth viewing experience. This results in a slightly delayed playback that becomes imperceptible once viewing has started.

The important point is that the Internet allows a choice of connectionless or connection-oriented transport protocols that designers can use to meet system requirements.

The Application Layer

The application layer is used for the actual data exchanges between processes running at endpoints, which is the ultimate purpose of remote connections. Everything else is, in a way, the mechanics of getting the data across. At this level, data are exchanged in a mutually agreed upon format with the shared semantics that the sender knows how to produce and the receiver how to interpret. This can be done through pairwise arrangements between communicating applications or with the aid of some common protocols and data definitions.

As Table 3.1 indicates, the Internet treats the application as a single layer, while OSI divides it in three: session, presentation, and application [2]. Although OSI is now mostly of historical interest, we briefly review the two lower levels of application layer to indicate the nature of anticipated design concerns at the application layer. The OSI session layer allows two communicating machines to establish a session between them. It provides some additional services, such as managing dialog control and direction of traffic flow, tokens, and synchronization that simplify coordination of communicating parties for protocols that require it. The presentation layer is concerned with the syntax and semantics of the transmitted data. It performs some functions expected to be fairly common that would otherwise need to be implemented individually within a number of applications. One of them is encoding of the data types and structures in a standard agreed upon way to bridge the individual differences between computers with different representations. The intent was to define data in a common abstract representation that different machines could translate to and from as necessary.

The Internet Protocol stack defines a single application layer. If some of the functionality provided by the OSI sub-layers is required, it needs to be provided by the communicating applications themselves or by the application protocols that they agree to use. Some commonly used Internet applications have evolved into using standardized protocols, such as HTTP (Hypertext Transfer Protocol) for the World Wide Web, SMTP (Simple Mail Transfer Protocol) and POP (Point of Presence Protocol) for email, FTP (File Transfer Protocol) for file transfers, and RTP (Real-Time Transport Protocol) for video over the Internet.

Data semantics, or the meaning of data, needs to be agreed upon by the applications either explicitly or through adherence to common specifications, such as HTML (Hypertext Markup Language) for text pages. Strictly speaking, HTML does not really define semantics of data but rather provides a convention for markup and encoding that indicates how text should be presented on the user interface to reflect the original design intent. Data semantics are in the displayed content itself, to be interpreted by human users possibly with the aid of a semantic definition such as a dictionary.

The challenge in IoT systems is for the semantics representation that facilitates "machine understanding" of data. The IIC IoT model [3] defines a framework at the application layer intended to foster connectivity and interoperability when the two communicating endpoints are part of an IoT system. Frameworks usually go beyond the cross-application connectivity by defining the syntax and semantics of data exchanges and possibly other elements of the common operating environment, such as naming, discovery, data exchange patterns, interaction APIs, and security. Such tight framework specifications are one way of dealing with the data semantics and interoperability problem in IoT systems. We discuss that problem, its common solutions, and some reference frameworks in chapters "IoT Data Models and Metadata" on data formats and "IoT Data Standards and Industry Specifications" on IoT standards.

IoT Network Layers

Figure 3.2 illustrates several concepts related of network layers of interest in IoT systems. It shows some practical combinations and variants of the layered design and protocol stacks introduced earlier. The physical layer illustrates several types of networks that may be encountered in IoT networks and on the Internet in general. They include telco cellular networks labeled as xG, such as GPRS, 3G, 4G, LTE, and 5G networks with their low-power, lower-bandwidth variants such as NB-IoT and LTE-M intended for use by sensors and things. Most of them are designed to support transfer of IP traffic above the link layer. The 802 series of standards define the PHY and often link layers for many networks that may be found in an IoT system, such as the IEEE 802.1 series of specifications for time-sensitive networks that facilitate deterministic delivery of packets required in many industrial uses. The IEEE series 802.3x are variants of Ethernet specifications, and the IEEE 802.11x

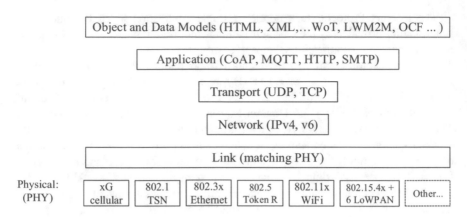

Fig. 3.2 Network layers in IoT systems

series cover Wi-Fi. The IEEE 802.15.4 series of specifications define low-power, short-distance (personal area) wireless networks that are of special interest for implementing sensor networks. When combined with the 6 LoWPAN adaptation layer, they provide IP compatibility at the network layer. All of them are discussed in more detail later in this chapter.

Figure 3.2 illustrates that individual link layers enable those networks to partici-pate and interoperate in Internetworking configurations by using the common net-working layer, the Internet Protocol (IP). IP has two compatible versions of addressing mechanism. The original IP definition is called v4 since it uses a 4-byte address field. The newer revision v6, discussed later in this chapter, uses larger address fields to accommodate the explosive growth in the number of endpoints largely brought about by IoT. The figure also indicates that most applications and services in IoT and on the Internet rely on UDP and TCP as transport protocols.

Internet Protocol layering is sometimes referred to as an hourglass or a narrow-waist protocol stack. This comes from the observation that there is a considerable variety of protocol layer implementations at the applications and link layers, but the IP layer with TCP and UDP has undergone a slow evolution since the Internet was invented. This is intentional, and it provides a stable base and continuing connectiv-ity for the existing Internet endpoints by not forcing major changes or replacements of the core infrastructure. The hourglass approach has proven its effectiveness, robustness, and scalability to handle the ever-increasing traffic volume and modali-ties, such as transmission of voice and video. At the same time, it provides ample opportunity for innovation and introduction of new types of networks below the IP layer (with their physical and link layers) and of application protocols above the transport.

Figure 3.2 illustrates several application protocols that may be encountered in IoT systems, including the CoAP (Constrained Application Protocol) which is designed for constrained devices and things, the MQTT messaging protocol, the web stalwart HTTP (Hypertext Transfer Protocol), and SMTP (Simple Mail Transfer Protocol) that may be used for operator messaging and notifications.

Inter-application data exchanges in IoT may require additional semantic annotation for proper processing by the receiving party. In addition to the Internet ubiquitous HTML and XML, Fig. 3.2 illustrates some IoT specific mechanisms that enable semantic interoperability, such as the WoT (Web of Things), LWM2M (Lightweight M2M), and OCF (Open Computing Foundation) framework, all of which are described later in chapter "IoT Data Standards and Industry Specifications" on IoT standards.

Edge Networks

Networks at the edge primarily connect sensors and things to each other and provide some means of connecting to the Internet, including the back-end part of the IoT system and the cloud.

If an IoT node has sufficient power and is within the reach of a cable or Wi-Fi connection to the Internet infrastructure, it can be fitted with the appropriate interfaces and protocol stacks and be ready to participate in data and command exchanges. The more difficult case of dealing with low-power, constrained, and dispersed sensor nodes and things is covered in the rest of this chapter. Before doing so, we briefly overview the common network topologies in the next section.

Network Topologies

This section provides a brief description of the more common types and basic principles of network topologies. They are applicable to all networks, but we primarily focus on what may be encountered at the edge of IoT systems. In practice, these networks may be combined using bridges and routers to form more complex systems with large number of nodes. The Internet is an extreme example of this, with billions of nodes on numerous networks and hundreds of thousands of routers in complex hierarchical arrangements of heterogeneous links and networks providing the overall connectivity.

There are numerous variations of network topologies. They represent logical or physical configurations of nodes, often depicted as circles, and links, commonly depicted as interconnecting lines. Physical topologies may reflect the physical layout of nodes and cabling. Logical topologies use the same symbols to depict which connections are possible between nodes that may differ from the physical connectivity given the node and protocol limitations.

Figure 3.3 illustrates some of the more common network topologies. Bus and ring topologies are used primarily in wired networks. They provide a shared communications medium that all nodes can access. Such networks and some variants of radio frequency selections form broadcast networks in the sense that all nodes can listen to all transmissions. Messages in a single transmission can be addressed to an

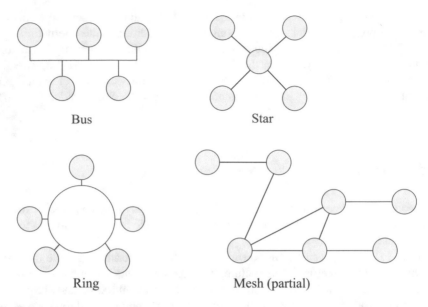

Fig. 3.3 Network topologies

individual node (unicast), all nodes (broadcast), and groups of nodes (multicast). Access to the shared medium requires arbitration that may be centralized in a segment-wide control point or distributed, where each node makes its own decisions in accordance with the common rules.

An example of the *bus topology* is the original Ethernet, where all nodes in a segment have access to a common shared medium and arbitrate access to it in a distributed manner. Access to the shared medium is governed by a Carrier Sense Multiple Access with a Collision Detection (CSMA/CD) scheme. When a node wishing to transmit detects an idle state of the carrier frequency on the shared link, it starts transmitting. All nodes listen to all transmissions to capture the data addressed to them. The transmitting node compares the bits on the line with its own transmission to check for discrepancies that indicate collisions occurring when two or more nodes attempt to transmit within the same time frame. Upon detecting a collision, each of the transmitting nodes backs out and retries after a period of time. The waiting times are somewhat randomized to avoid repeated collisions that would result from synchronized retransmission attempts. One of the problems of bus topologies with shared access and equal priority is the potential for unpredictable and variable delays in successful message transfers due to collisions.

Contemporary implementations of Ethernet use more elaborate forms of structured wiring and point-to-point connections to switches and hubs organized in a tree-like physical topology. This simplifies wiring, increases network capacity in terms of the number of nodes that can be supported, and tends to alleviate but does not eliminate the problem of variable delays.

Ring topologies basically operate by circulating messages across links that form a physical or a logical ring. Upon gaining access, a transmitting node places its message on the ring. All nodes can monitor traffic on the ring. The intended destination node receives the message and typically removes it from further circulation. A variant, called token ring, uses a special type of a circulating control message – called the *token* – as a permission to transmit. A node wishing to transmit waits for the receipt of a token, removes it from the ring, and places its payload-bearing message in the next available network slot. Following its message transmission, the node holding the token passes it along. Use of tokens as permissions to send eliminates collisions and can increase the effective throughput of the network. It also bounds delays and reduces jitter by limiting the maximum wait time to a full token circulation of the ring. Multiple tokens may be used and prioritized to give preferential treatment and tighter guarantees to the time-critical traffic. Since existence of at least one token is essential for operation of the ring, special provisions – usually in hardware – need to be made to preserve token integrity and to regenerate it if necessary when nodes fail or are powered off. One of the successful commercial implementations of this approach was the mostly proprietary IBM Token Ring LAN. It had 16 Mbps signaling speed which competed favorably with its then contemporary 10 Mbps Ethernet in terms of throughput and the ability to limit delays and jitter and thus carry multimedia traffic. Ethernet won by being a more open specification with multi-vendor support and relatively quickly advancing to 100 Mbps and later to 1 Gbps speeds. As is often the case in the computer industry, openness fosters competition which contributes to a wider availability of components, increases volumes, and lowers prices that in turn lead to greater adoption.

Other type of networks, such as star and mesh in Fig. 3.3, have point-to-point links. *Star* is a physical or logical topology with a single node in the center and leaf nodes connected to it via direct links. Star topologies are used in some sensor networks, in telco networks, and in structured variants of wireless networks such as Wi-Fi where nodes communicate directly with the access point.

A *mesh topology* refers to various combinations of nodes with some number of communication links. In a fully connected mesh, each node is connected to every other node. In a partial mesh, such as the one shown in Fig. 3.2, there is a subset of links representing direct connectivity between a subset of nodes. Such networks are multi-hop in the sense that some messages may have to traverse more than one intermediate node and links before reaching the intended destination. Multi-hop networks require use of routing mechanisms to determine message traversal paths.

Wireless Edge Networks

In the remainder of this chapter, we focus primarily on wireless edge networks that are suitable for IoT uses. They may be used in a variety of settings with no networking infrastructure, such as to monitor soil and ambient conditions in agricultural fields with battery-powered sensors or in geographically dispersed and/or remote locations such as open-pit mines and oil and gas pipelines.

The selection of networks described in this chapter is governed by their potential significance to IoT, either by virtue of suitability, adoption, or installed base. We describe their basic operation, capabilities, and limitations to aid in understanding of their role in the overall IoT system. Our coverage is focused on principles and mechanisms that may be of general interest in IoT system design. This is not intended to be an exhaustive survey or a detailed description of operation of specific networks.

The "no new wires" also makes wireless networks attractive in any setting where wired installation of sensors and things is not feasible or practical. This may include home automation, patient monitoring, industrial applications requiring addition of new monitoring sensors, automotive, and virtually any other application where mobility is necessary or desired. In addition to the lower cost, lack of wires makes wireless sensor installation quick and comparatively painless thus making them attractive even for temporary uses, such as leak detection in new factories, with the potential for later reuse elsewhere.

While wireless networks may be used to create self-contained operational enclaves, edge networks in an IoT system have to be bridged to the Internet, usually via edge gateways or routers. Depending on the topology and availability, those edge routers may use log-range wireless or wired back-haul connections to the Internet themselves. With Internet connectivity being the goal for IoT nodes directly or indirectly, it is generally beneficial to use edge networks and protocol layers that are Internet-compatible or relatively easy to convert to. Some of the networks described in this chapter are designed with that objective in mind. A few are not, but we include them because they have a significant installed base of devices at the edges that may benefit from being brought to the IoT fold.

Some characteristics of interest when evaluating wireless networks include:

- Range
- Bandwidth
- Topologies
- Determinism – latency and jitter
- Energy usage and power requirements
- Frequencies

Range indicates the operational distance between nodes that the network is designed to support. A coarse classification is into short, medium, and long. Short is generally used as a wireless replacement for serial cable and defined as being on the order of 10 m between devices. Medium range usually refers to distances of hundreds of meters but less than a kilometer. Long range typically refers to distances of over a kilometer (or mile in some definitions). These distances refer to what a given technology can support. In actual deployments, a network's effective operating distance may be shorter due to obstacles and interference that may be caused by a number of factors, including density and contention in the chosen area or adjacent frequencies, electromagnetic interference from other sources such as appliances and automated doors, weather, and physical barriers such as walls.

Bandwidth is usually specified as the nominal signaling rate of a network. This is basically the upper speed limit that cannot be exceeded, but the actual data rates experienced by the two communicating applications are usually considerably lower. This is due to a number of factors, including sub-optimal operating conditions, network contention and retry delays, and lost or corrupted messages. Effective throughput is also adversely impacted by the overhead of executing software that implements networking layer stacks at both origin and destination. It is not uncommon for the effective throughput experienced between the two communicating applications, sometimes called "goodput," to be reduced to a fraction of the nominal signaling rate.

Determinism of packet deliveries is a function of delay and jitter, which is variance in delay. They are determined by a number of factors that can impact the maximum and average delays encountered in successfully delivering a message, such as network efficiency, arbitration mechanism, and number of hops that may need to be traveled.

Energy usage is primarily determined by the power requirements of the transceiver in the transmit and receive modes and their duty cycles, i.e., on and off times. Wireless networks usually specify minimums required to support the network and the intended range. Maximum power and duty cycles are often limited by the regulatory agencies and enforced by requiring certification of compliance.

Frequencies and radio spectrum allocation are discussed in the next section.

Radio Spectrum and Frequency Allocation

Wireless communication requires the use of radio transceivers. The choice of transmitting and receiving frequencies is governed by the cost and performance considerations, and it is often limited by regulation. The radio spectrum has many users, and governments and regulatory bodies divide and manage it in two broad categories – *licensed* and *unlicensed*. The licensed part of the spectrum is allocated for specific uses, and parts of it are assigned or sold to commercial licensees. Some uses of the licensed spectrum include radio and TV broadcasting, aeronautical and marine navigation and communication, emergency responders, and other official communications. Commercial telco operators are significant users of the licensed spectrum for mobile communication and data services. As the name implies, use of any portion of the licensed spectrum requires a license to operate that usually needs to be paid for.

Unlicensed portions of the radio spectrum may be used without license or charge to operate. However, they can still be regulated in terms of signal power and the duty cycle or the permitted duration of a continuous active transmission. Commercial devices and transceivers may need to be certified by authorized bodies to meet these requirements. Parts of the unlicensed spectrum are commonly reserved for Industrial, Scientific, and Medical bands (ISM). Details of spectrum allocation and operating limits tend to be specific to geographies and countries. An IoT system intended for international use will need to meet the specific local requirements in the areas where it is installed.

The most commonly used unlicensed frequency bands for IoT are in the 2.4 GHz range that is used by IEEE 802.11 Wi-Fi networks, Bluetooth, and IEEE 802.15.4 WPANs. Sub-GHz bands are also used in the ISM range of 868 and 915 MHz with some variations worldwide. In general, sub-GHz frequencies have better signal propagation characteristics and thus longer effective operating range and distance between devices. On the negative side, they support lower signaling rates which may not be an issue for some IoT sensor applications.

Form the practical point of view, IoT systems may use unlicensed or licensed spectrum. The licensed spectrum is typically used by purchasing a service from a commercial operator who has the right to use it, such as a telco operator. This would normally be required to use cellular or mobile services, such as variants of GPRS, LTE, and 5G.

Short-distance and some medium-distance wireless networks for IoT commonly use unlicensed portions of the spectrum. The advantage is no licensing cost, but the drawback is that other networks and users of the spectrum may be encountered in the operating area and adversely impact performance and usability. Even for short-distance uses, it is a good practice to conduct a radio frequency survey of the planned deployment area to make an informed selection. This may also include an analysis of fading and multipath interference likely to be encountered due to natural and man-made barriers in the area of operation.

IEEE 802.11, Wi-Fi

IEEE 802.11 networks, better known by their commercial designation Wi-Fi, are being used by more devices than there are people on our planet. They are widely available on smart phones, laptops, notepads, smart thermostats, smart speakers, security cameras, and numerous other devices, thus making them a default go-to candidate for wireless IoT uses in consumer and commercial settings. Due to their power requirements, Wi-Fi networks are used by nodes with mains power supplies or bigger and easily recharged batteries. They are not suitable for constrained low-power devices.

The IEEE 802.11 set of specifications define the PHY and MAC used by the Wi-Fi networks. They are designed for short-range use of up to about 50 m indoors with longer distances possible outdoors, subject to power, interference, and obstacles that include topography and weather. The Wi-Fi primary focus is on achieving high data rates and not power conservation, thus making it suitable for power-affluent nodes within the range of an access point (AP or WAP).

IEEE 802.11 networks operate in unlicensed ISM radio frequencies primarily centered around 2.4 and 5 GHz. Since the first specification in 1997, a series of specifications and amendments, designated by one or two lowercase trailing letters, such as 802.11g and 802.11ad, were issued to address different speeds using various forms of modulation, frequencies, and antennas. Some more recent specifications also define operations in sub-GHz range for better propagation and in the

60 GHz range with higher speeds. Over time, the speeds have increased from 1 Mbps to 6, 9, 12, 18, 24, 36, 48, and 54 Mbps and to 10s of Gbps in the more recent specifications.

Constrained Devices and Networks

Edge networks described in this section are used by things and sensors at the edge that do not have the capability to connect directly to the Internet. This can be caused by remote placement away from the established Internet infrastructure or by a number of other constraints in the operating environment, the edge networks, or the devices themselves.

Constraints determine the limitations of the functional capabilities of things at the edge. A more formal classification of device constraints in terms of their hardware configurations is listed in Table 3.2 [4]. The table specifies representative data and program memory sizes and not the processing power. However, for this class of embedded microcontrollers and SoC devices, the processing power is usually matched accordingly. The specific numbers, such as those in Table 3.2, should be used for illustrative purposes only as they tend to become obsolete rather quickly with changes in technology. However, the functionality description is a good starting point for longer lasting classification purposes.

Constrained networks are sometimes characterized as Low-Power and Lossy Networks (LLNs). They are primarily intended for use by the constrained devices with processing and power supply limitations. LLNs themselves can be constrained due to several reasons, including cost constraints, node constraints, power and environmental constraints, limited spectrum, high density, interference, and the like.

One of the key considerations in the design and implementation of LLN networks is to conserve power. Some of the techniques commonly used to achieve that purpose include:

Table 3.2 Constrained devices

Name	Data size, e.g., RAM	Code size, e.g., flash	Functionality
Class 0, C0	≪10 KB	≪100 KB	Very constrained devices, e.g., sensor-like motes. Cannot communicate with the Internet directly, need help of larger proxies, gateways, or servers
Class 1, C1	~10 KB	~100 KB	IP and security capable, cannot easily communicate using full Internet Protocol stacks, such as HTTPS. May be able to use CoAP over UDP; can be integrated as peer nodes into larger IP networks
Class 2, C2	~50 KB	~250 KB	Support most of protocol stacks used on notebooks and servers; constrained, can benefit from lightweight and energy-efficient protocols and from consuming less bandwidth

- Communication and mediation schemes that allow nodes to sleep
- Smaller (compressed) headers
- Smaller payloads/packets to keep bit error rates low and permit media sharing
- Limited bandwidth
- Nodes with different capability, some with reduced functionality and dependence on more powerful peers

Figure 3.4 illustrates a generic example of a partial mesh edge network of LLN type. One of the techniques commonly used in such networks is to intersperse nodes of different capability throughout the network. The more capable nodes, labeled as full-function (FF) nodes in Fig. 3.4, provide the more stable part of the network structure. They are usually comparatively power-affluent, can connect to more than one node, and participate in mesh routing of packets between nodes that are not directly connected. This allows incorporation of highly constrained network nodes, marked as reduced-function (RF) nodes in Fig. 3.4. They usually have very limited power and processing resources and occasionally go to very low-power sleep states with no or limited connectivity to conserve energy. Such nodes are usually able to connect only to higher-power nodes and not to each other directly. The dashed lines on their communication links in Fig. 3.4 indicate intermittency of what are sometimes referred to as "sleepy links." Judicious mixing of diverse types of nodes in LLNs provides the environment and support for highly constrained nodes to operate in a more power frugal manner.

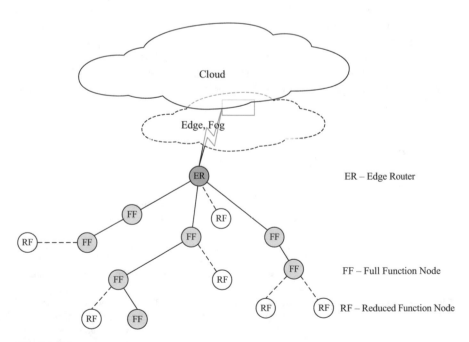

Fig. 3.4 Edge network

In addition, at least one highly capable node in the network functions as the border or the edge router (ER) that connects the entire mesh to the next-level network, usually the Internet. Such nodes are often in charge of performing other important functions, such as forming and managing the network and its operation. The edge router can be a stand-alone node, or it can be integrated in an IoT gateway.

As described in later sections, different specifications use different names for nodes with different capabilities and functions, such as router or FFD for full-function devices, end node or host for the reduced-function ones, and the like. However, the concept of mixing nodes with different capabilities within a network is the same.

IEEE 802.15.4

One of IEEE working groups is the 802.15 for Wireless Personal Area Networks (WPANs) and its subgroup described in this section, IEEE 802.15.4 called Low-Rate Wireless Personal Area Network (LR-WPAN or LoWPAN). Personal designation in these names is somewhat of a legacy moniker that results from the original assumption that (short-range) wireless networks would be used primarily to connect personal devices in the proximity of a user. In IoT, they are mostly used to connect constrained things and sensors.

IEEE 802.15.4 was originally ratified by IEEE in 2003 [5], and a number of subsequent additions and modifications were added to handle spectrum needs of different geographies and uses, such as the smart grid. Its variants are usually denoted by appending a designation letter, such as IEEE 802.15.4a, c, d, e, f, and g.

IEEE 802.15.4 is intended for use over distances of about 10 m with data transfer rates of up to 250 Kbps. Its focus is on low-cost and low-power requirements for communication among nearby devices with little or no underlying infrastructure. It is applicable for use by remote sensors that may be unattended and need to operate on battery power for years.

IEEE 802.15.4 is a standard designed to provide a framework for implementation of low-cost wireless networks. It only provides a definition of the two lower network layers, PHY and MAC. Specification of the upper layers is left out for specific application and uses to define. When implemented on constrained nodes, these are usually less complex and resource-intensive protocols than the Internet layer stack described earlier in this chapter. Some of them, such as ZigBee, Thread, 6 LoWPAN, WirelessHART, and ISA 100.11a, are described later in this chapter.

One of the benefits of standardizing lower layers is that many different types of networks and protocols can be built on essentially the same hardware foundation of transceivers and network interfaces. Microcontrollers and SoCs intended for use in 802.15.4 compliant networks can implement PHY and all or some of the MAC functions using common hardware modules. Implementation of the chosen upper layer can be in software or in firmware, and its loading can be deferred until manufacturing or even deployment time. This commonality increases attractiveness to hardware manufacturers of network components by increasing potential sales volumes. Existence of low-cost hardware in turn facilitates construction and deployment of IoT sensors and things that use such networks.

IEEE 802.15.4 Physical/Link Layer

IEEE 802.15.4 is designed to operate on unlicensed spectrum, including sub-gigahertz bands such as in the 868.0–868.3 MHz range for use in Europe, 902–928 MHz for North America, and 950–956 MHz in Japan. Gigahertz frequencies in the 2400–2483.5 MHz range are available for use worldwide. Depending on the frequency and power levels of the radio transceivers used, the number of usable channels in a network can vary from 1 to 16 and even 30, with bit rates of 20, 40, 100, and 250 Mbps.

The chosen communication frequency is shared by all nodes in the network. Since radio networks are usually half-duplex meaning that a node cannot listen to its transmission on the link, access to the shared medium is governed by the CSMA/CA (Carrier Sense Multiple Access/Collision Avoidance) technique conceptually similar to the one described in conjunction with the Ethernet earlier in this chapter. This means that nodes wishing to transmit will try to do so when detecting an idle line. If a collision occurs, the node will back off for a random period of time – calculated using the algorithm in IEEE 802.15.4 specification – and try again. As described later, IEEE 802.15.4 also defines the use of optional beaconing that may be used to guarantee conflict-free transmissions within a super frame.

As is common in LLNs, each node in a network may be a full-function device (FFD) or a reduced-function device (RFD). Each network needs to have a coordinator, an FFD device that is in charge of the entire network. It is usually called a PAN coordinator for Personal Area Network coordinator. An FFD can participate in a network of any topology, implement the complete protocol set, talk to any other device, and be a PAN coordinator. Reduced-function RFD node type is limited to star topology or to being an end node in a P2P network. Such a node usually has a simple implementation with potentially reduced protocol stack, and it cannot become a PAN coordinator.

Network topology can be a star or a point-to-point, peer-to-peer (P2P) network with a mix of FFDs and RFDs in a variety of configurations. Since RFDs can only be leaf nodes that can communicate with FFDs, a simple star configuration consists of a single PAN (FFD) node in the logical center and a number of leaf RFD nodes. The PAN would then connect to the Internet directly or via a gateway to relay the relevant IoT traffic. Since the IEEE 802.15.4 standard does not define functions above the OSI layer 2, it does not include routing, and networks built on top of it can supply their own solutions to support a variety of topologies.

The PHY layer provides the data transmission via modulation and bit encoding over the shared link. It managers the radio frequency (RF) transceiver, including channel selection within the chosen frequency band and handles energy selection and management functions.

The MAC (link) layer enables transmission of frames over the physical channel. On the receiving end, it performs frame validation and error detection. It also handles collision avoidance and provides access (hook) points for security services. In addition, it manages network beaconing. Two successive beacons are used to delineate what is called a super-frame structure that consists of 16 equal length time slots, up to 7 of which can be reserved for use by the specific applications in contention-less manner. In beaconless networks, there are no reserved time slots, and nodes use the CSMA/CA protocol to manage contention. Maximum payload size per packet in IEEE 802.15.4 networks is 127 bytes.

In IEEE 802.15.4, messages may be acknowledged if requested or unacknowledged. They may be sent directly from the originator to the recipient or sent indirectly via the coordinator.

Amendments of the IEEE 802.15.4 specification provide additional enhancements, such as the more power-efficient time-slotted channel hopping (TSCH) and coordinated sampled listening (CSL) in 802.15.4d and PHY layer suited to the increased range in sub-GHz networks. The IEEE 802.15.4e defined MAC amendments to support channel hopping to increase robustness against interference and multipath fading that are important in industrial applications.

The IEEE 802.15.4 MAC layer includes provisions for implementing security by higher layers. They may specify keys for symmetric cryptography to encrypt transmissions or restrict it to groups or just pairs of devices. Access Control Lists (ACLs) stored at devices can be used to identify trusted neighbors and security methods to use.

6 LoWPAN

The key components for enabling pervasive sensing of the physical world in IoT systems are sensors and things connected by the low-power wireless networks in geographically dispersed locations with little or no networking infrastructure. As indicated, such devices and networks may have a number of constraints that dictate design compromises and implementation restrictions. Their primary function is to capture real-world data for processing by IoT systems whose major portions reside on or are directly connected to the Internet. Enabling such devices and networks to effectively participate in the Internet-style Internetworking would obviously realize a number of advantages and efficiencies in the overall IoT system design. The 6 LoWPAN is a standardized protocol adaptation layer that accomplishes this purpose.

6 LoWPAN is an acronym for IPv6 over Low-Power Wireless Personal Area Networks. It defines the use of IP networking over IEEE 802.15.4 low-power wireless networks [6]. 6 LoWPAN brings the IP capability to edge devices on low-power wireless networks so that they can become sources and destinations directly reachable from the Internet. It makes message exchanges within the low-power wireless segments compatible with the Internet to enable transitions between the two that can, in effect, achieve global connectivity.

The use of IP provides a number of benefits, including interoperability with all other network IP links, established naming and address translation, established transport and application protocols, scalability, security, reliability, and manageability.

The use of IPv6 provides ample addressing capability to accommodate projected growth of IoT endpoints well into the future. IPv4 has a 32-bit address filed thus setting an upper limit of 2^{32} distinct addresses or approximately 4 billion. Some of those addresses are reserved, so the actual usable number is quite a bit lower. As the number of Internet devices increased with the growth of web hosts, personal computers, and mobile phones, IPv4 started approaching its limit in terms of its ability to uniquely address everything connected to the Internet. The problem is somewhat alleviated with network address translators and dynamic assignment of temporary IP addresses on network segments, such as LANs and Wi-Fi zones. However, the number is clearly inadequate for the billions of IP addressable IoT things that are on the horizon. IP version 6, or IPv6, has the 128-bit address field and thus the theoretical capability of undecillion unique addresses, which is a number on the order of 10^{36}, or a trillion (10^{12}) cubed. Again, some addresses are reserved, but the magnitude of the number is immense and sufficient for the projected IoT needs with room for considerable expansion.

6 LoWPAN is the network-layer building block that brings Internet compatibility via adaptation to constrained nodes and networks. It builds on top of the IEEE 802.15.4 PHY and MAC and works well with CoAP, described later in this chapter, to complete the protocol stack suitable for use with constrained IoT devices and networks.

In terms of implementation, the 6 LoWPAN is an adaptation layer that sits above the IEEE 802.15.4 MAC layer and below the networking layer, as shown in Fig. 3.5. It enables constrained networks that implement it to complete the IP stack and thus interoperate at the Internet networking layer. Since only IPv6 is supported, ultimate endpoints need to use that form of addressing. Some legacy network implementations may contain parts that support only IPv4. This does not

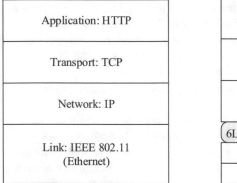
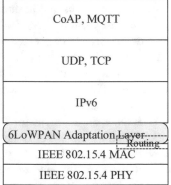
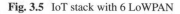

Fig. 3.5 IoT stack with 6 LoWPAN

restrict the use of 6 LoWPAN, since there are well-known methods for tunneling IPv6 over IPv4 that are transparent to endpoints.

6 LoWPAN provides a definition for encapsulating UDP and even TCP packets into payload of IEEE 802.15.4 MAC frames. The specification accommodates LLN limitations and needs of constrained devices by compressing headers and addresses which reduces overhead and allows more room for payload. This is made possible by using defaults and shortened addresses for intrasegment communications among nodes that are within the same PAN. 6 LoWPAN uses some clever tricks to reduce IP packet header fields, such as shortening the source and destination addresses and length fields in packet headers and deriving them from the link layer addresses [7].

In the best-case scenario of a UDP packet transmission, 6 LoWPAN can compress a 40 byte IPv6 address and 8 byte UDP header to only 6 bytes, thus extending the available payload to 108 bytes of the maximum of 127 bytes that IEEE 802.15.14 MAC frame allows. In comparison, uncompressed IPv6 address and UDP header would leave a maximum of 53 bytes of the actual data payload. However, for packets destined for segments of the fully functional Internet, header and address compression may need to be eliminated or reduced only to the extent that allows edge router nodes to reconstruct the omitted pieces.

Since IPv6 allows a Maximum Transmission Unit (MTU) length of 1280 bytes and the IEEE 802.15.4 limit is 127, 6 LoWPAN supports segmentation and reassembly of packets to handle longer payloads. Naturally, an efficient practical alternative is to limit the payload size in the application protocols when communicating with known constrained endpoints.

In order to support mesh configurations that IEEE 802.15.4 allows, 6 LoWPAN headers make provisions for mesh routing. This is different from the regular IP routing and is sometimes referred to as the *mesh-under routing* (path computation and forwarding) that operates at the adaptation layer and within the limits of a 6 LoWPAN network and is thus hidden from the rest of the IP layer. An alternative is to use the Internet IP layer network routing, sometimes referred to as the *over-mesh routing*.

There are three types of nodes in 6 LoWPAN networks: edge router, router, and host nodes. Their key functions are:

- Edge or border router (LBR) – connects 6 LoWPAN network to the Internet as a fully functional IP networking layer node and router. It performs three functions: (1) data exchange between 6 LoWPAN nodes, Internet, and other IPv6 networks, (2) data exchange between nodes within the 6 LoWPAN network, and (3) generation and maintenance of the 6 LoWPAN radio network.
- Router node – route messages between nodes within the 6 LoWPAN network in mesh configurations, cannot sleep.
- Host node – end nodes, cannot route traffic, can sleep periodically to conserve power.

A border router is the proper IP router that connects 6 LoWPAN networks to the Internet and other IPv6 networks such as 6 LoWPAN clusters. It can handle header adjustments, addressing, and routing as necessary for traffic flows from the Internet to 6 LoWPAN local nodes in its sub-network, traffic from local nodes to the Internet, and intra-6 LoWPAN traffic between local nodes. Border router can be physically connected to the Internet via whatever suitable links may be available, such as LAN to a local access point or telco or WAN connection.

Router nodes provide the routing tables and forwarding functions for multi-hop communication between the local nodes in a mesh. They are roughly equivalent to an intra-PAN full-function device. Host nodes in 6 LoWPAN are effectively reduced-function devices that can only talk via direct links to the more capable nodes. The most constrained devices in 6 LoWPAN networks are implemented as host nodes. As is common in LLNs, these devices can enter very low-power modes in which activities and communications are suspended for periods of time to conserve power. The power and value of 6 LoWPAN networks are that even such nodes can be addressed and exchange data with other entities on the Internet, with the 6 LoWPAN infrastructure and more capable nodes aiding by managing routing and header expansion and compression but without altering the payload. Given that the ultimate destination of IoT data is often some aggregation and processing point on the Internet with commands flowing in the opposite direction towards the edge actuators, this is a unique advantage of 6 LoWPAN in comparison with non-Internet-compatible systems with proprietary stacks such as ZigBee and Bluetooth described later. As is the case with the network-layer IP routers, 6 LoWPAN routers are stateless in the sense that they do not maintain any application-layer state. In contrast, other proprietary systems usually need to deploy more complex and state-aware translations between the Internet and their own internal application domains.

Routing in 6 LoWPAN mesh configurations is commonly implemented using the IPv6 Routing for Low-Power and Lossy Networks (RPL) (AN ref RFC 6550) [8]. It is a distance-vector routing protocol that determines paths by constructing and maintaining Destination-Oriented Directed Acyclic Graph (DODAG). Such a graph has no cycles (message loops) and is directed towards a specific node (destination). The protocol is designed to be adaptive and to reconfigure the routing tables as necessary when nodes leave or enter the network.

6 LoWPAN specifications also include definition of neighbor node discovery and network formation, including the ability of nodes to derive their own IPv6 subnet addresses. Their details are out of scope of this book.

IEEE 802.15.4 IP-Based Networks

There are several protocols and types of LLNs that are based on IEEE 802.15.4. Some of them that are of interest for IoT applications and are IP-based are briefly described in this section.

Thread

Thread is a networking protocol stack designed for home automation. It is intended to provide secure reliable mesh networking for IoT with no single (device) point of failure, with simple connectivity, and with low power. It provides mechanisms for simple network installation and start-up using smart phones or computers. It incorporates IPv6 addressability and uses 6 LoWPAN at the network layer thus requiring a border router. Thread uses UDP at the transport layer and AES encryption for security.

Thread focuses on networking and does not define an application layer. Those could be provided by other frameworks, such as ZigBee and OCF.

Thread was created by a group of companies including Nest (Google), ARM, Qualcomm, Samsung, several silicon manufacturers, and home devices such as Osram lighting and Yale locks and later joined by Apple [9]. The Thread Group manages technical specifications and provides product certification to assure compliance.

WirelessHART

WirelessHART is the protocol stack intended for process control applications. Its origins are in the wired HART (Highway Addressable Remote Transducer Points) used in industrial control [10]. WirelessHART uses the IEEE 802.15.4 PHY foundation to create a wireless version. It supports time-synchronized, self-organizing, and self-healing star and mesh network configurations. It uses 2.4 GHz ISM band and provides data rates of up to 250 Kbps with half-duplex radio transceivers ranging up to 100 m line of sight.

The network uses channel (frequency) hopping to avoid interference. Transmission time slots are synchronized, and they may be used as dedicated or contention-based, with collision avoidance and backoff. It uses AES-128 encryption for security.

ISA100.11a

ISA 100.11a is a standard defined by the Industrial Society for Automation (ISA) [11]. It is intended to provide reliable and secure wireless operation for non-critical monitoring, alerting, supervisory control, open-loop control, and closed-loop control applications. It defines the protocol suite, system management, gateway, and security specifications for low data rate wireless connectivity with fixed, portable, and moving devices supporting very limited power consumption requirements. The application focus is to address the performance needs of applications such as monitoring and process control where latencies on the order of 100 ms can be tolerated, with optional behavior for shorter latency.

Its network layer is based on 6 LoWPAN IPv6 with UDP or TCP at the transport layer. Security is based on AES-128 bit encryption.

Non-IP IoT Wireless Edge Networks

This section overviews several wireless edge networks that are not IP-compatible but have a sizable installed base of things and smart devices. This makes them likely to be a consideration in designing and integrating IoT systems.

ZigBee

ZigBee is a protocol specification created and maintained by an industry alliance. It uses IEEE 802.15.4-based PHY and MAC layers and builds its own network and application layers above them (AN ref ZigBee) [12]. The alliance issues specifications and provides certification of product compliance for cross-vendor interoperability. It currently claims over 400 member companies and more than 2000 certified products used in a variety of applications including home automation, appliances, smart meters, retail, lighting, medical devices, and others.

ZigBee is designed to support collections of devices in relatively close proximity, including ones that are battery operated, by means of a low-cost, low-power, low data rate network. The idea was conceived in 1998; the first specification was released in 2004 with multiple subsequent releases and changes with enhancements and repositioning towards IoT. Since each specification has a compliant base of installed products, an effort is made to provide for backward compatibility with earlier versions.

Technically, ZigBee is a self-healing, self-forming network that supports star and mesh configurations. It uses the 2.4 GHz global unlicensed band with support for 16 channels of 2 MHz band therein to avoid collisions with Wi-Fi installations. The rated speed is 250 Kbps at distances of 10–100 m, depending on the power output and environmental characteristics. Sub-GHz bands are also supported in the 868 and 915–921 MHz range to accommodate international requirements. Use of those frequencies reduces speed to 100 and 10 Kbps, respectively, while increasing the line of sight range to up to 1 km. Security is supported at both network and application layers.

ZigBee nodes may be one of three types:

- Coordinator (C) – one per network, never sleeps (mains powered), starts a new PAN, selects the PAN ID and operational channel, can assist in routing data in mesh configurations, allows R and E devices to join network
- Router (R) – can route traffic in mesh configurations, after joining PAN can allow other R and E devices to join, can support child devices, never sleeps
- End device (E) – battery-powered, can sleep to reduce power consumption, can talk only to a parent node C or R, cannot support child nodes

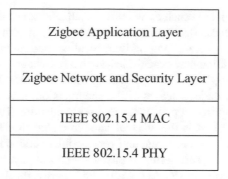

| Zigbee Application Layer |
| Zigbee Network and Security Layer |
| IEEE 802.15.4 MAC |
| IEEE 802.15.4 PHY |

Fig. 3.6 ZigBee protocol stack

Router nodes are primarily used in mesh configurations to extend range by connecting segments. The routing protocol used by ZigBee is called Ad hoc On-Demand Distance Vector (AODV) [13]. It is a reactive protocol that determines a route when a message needs to be sent. As a result, ZigBee mesh networks are reconfigurable and self-healing in the sense that new traffic will bypass failed or disconnected nodes.

As illustrated in Fig. 3.6, ZigBee provides its own network and application layers on top of the IEEE 802.15.4 foundation. The networking layer manages network functions, such as formation and routing. It also supports the use of 128-bit Advanced Encryption Standard (AES) security for transfer of encrypted messages. Encryption keys can be assigned to a network or to a link.

There is also an IP version of the ZigBee stack that uses 6 LoWPAN described later in this chapter; however, it works only with select flavors of the application layer.

The ZigBee application layer provides several flavors of a fairly elaborate framework functions. They support cross-application interoperability by defining profiles, called clusters, of commonly used objects and their properties. Clusters exist for measurement of temperature, flow, pressure, humidity, occupancy, HVAC, lighting, and lighting control among others. Their use facilitates interoperability within ZigBee compliant devices and applications.

Bluetooth

Bluetooth is a wireless connection technology originally devised by an industry consortium to replace wires with radio links for connecting PC peripherals [14]. It is designed for low-power, low-cost, short-range applications. It is not IP compatible, and it does not use the IEEE 802.15.4 PHY and MAC layers. However, Bluetooth has significant market acceptance and great importance for IoT with its installed base of over 4 billion devices. Its most significant market driver is the inclusion in smart phones and personal devices. Major uses of Bluetooth include audio streaming devices – such as wireless headphones, speakers, and in-car systems – sports and fitness (trackers and smart watches), health and wellness, home automation, PC peripherals, and indoor location and asset tracking systems. The association is adding new capabilities, such as extended range and mesh topology, to address the needs of industrial control and automation systems.

There are several major releases of the specification, including Bluetooth Low Energy (BLE), that differ in power, range, and bandwidth. Two major variants of Bluetooth exist. The first is the original version called the Basic Rate/Enhanced Data Rate (BR/EDR). A more recent and very popular version is the Bluetooth Low Energy (BLE). They differ in terms of physical and link layers. Peripheral devices may implement only one version to save cost and energy; this is most commonly the LE version. More power-affluent nodes, such as mobile phones, usually implement both to be able to communicate with a wider assortment of devices, including wireless headphones that commonly use BR/EDR for its higher data rate. Communication is usually structured in a master-slave fashion. A master node can discover and support thousands of devices in secured and authenticated exchanges.

Depending on the transmitter power, Bluetooth range varies from 0.5 to 100 m or longer with the 10 m nominal being most commonly used. Bandwidth ranges between 1 and 3 Mbps. Bluetooth operates using the unlicensed ISM frequency generally in the 2.4 GHz range. Most commonly used topologies are point-to-point and star, with the mesh being introduced recently.

The Bluetooth LE frequency band is subdivided in 40 channels, 3 of which can be used for advertising and 37 for data. The system uses adaptive frequency hopping among channels to select the ones with better signal quality and to avoid interference. Advertising channels carry advertising packets that may be used for discovery and broadcasting. Data channels carry data packets for exchange of data between applications. Channel selection is flexible and negotiable to avoid interference with other sources that may be using the same 2.4 MHz band, such as Wi-Fi.

Bluetooth is designed to facilitate a relatively opportunistic discovery and communication among devices in ad hoc arrangements as well as the more efficient modality in known permanent configurations. Devices that have affinity for each other, such as the same owner, can be intentionally paired to allow direct communication with reduced overhead. For example, with your permission, your fitness tracker may pair with your phone, and your phone may pair with your car's entertainment system.

A Bluetooth generic access profile defines four roles that endpoints can assume over a particular link:

- Broadcaster – can send advertising events, has a transmitter, receiver is optional
- Observer – receives advertising events, has a receiver, transmitter is optional
- Peripheral – can accept connections, when connected operates in a peripheral (slave) mode, has a transmitter and a receiver
- Central – can initiate establishment of connections, when connected operates in central (master) mode, has a transmitter and a receiver

Nodes can change and assume multiple roles in different connections as necessary. Packets with advertising events can contain additional data that describe the advertise-

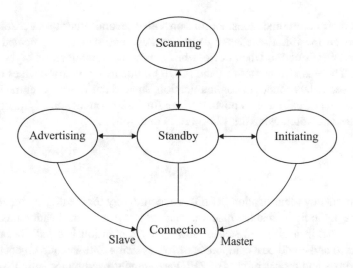

Fig. 3.7 Bluetooth LE states and transitions

ment, such as a desire to connect or data for observer nodes to consume. Advertisements can be general, intended for any receiver in the range, or directed. Directed advertisements are addressed to a specific peer, such as a previously paired node, inviting it to connect as fast as possible. For notifications, a retailer may place Bluetooth-based beacons in a store and send advertisements that your phone can receive when you are nearby, without necessarily engaging in follow-up communications.

Figure 3.7 illustrates a BLE link-level state diagram. We provide a somewhat simplified description of sequences of events to illustrate key activities and transitions. An observer node can be in the scanning mode to listen to general advertisements, or in the standby mode to conserve power. Similarly, beacon (or broadcaster) nodes can oscillate between advertising and standby states. A peripheral wishing to establish a connection enters advertising mode and sends advertising packets to announce its intent. A central node potentially interested in the connection can request more information and initiate connection, acting as a master. Upon receiving acknowledgment, a peripheral node enters connection as a slave. Once connected, the two can exchange messages in either direction.

Bluetooth is not IP compatible, and it provides its own networking and application stacks. It includes a logical link control and an Adaptation Protocol (L2CAP) that handles link state management and transitions. There is also a specification for carrying IPv6 traffic over BLE version 3.1 and higher using techniques similar to 6 LoWPAN [15].

Bluetooth stack also contains attribute and access profiles that define general types and characteristics of devices to simplify application exchanges through a shared understanding provided by the specification. There are also device profiles for the specific domains. For example, the health device profile includes definitions of the common device types, such as blood pressure monitor and body thermometer.

One of the functional components in the Bluetooth protocol layer stack is the security manager. It manages and authenticates bonding of devices via pairing and encrypts their communications. Pairing provides several security mechanisms that are based on the initial exchange of public keys and can be augmented by user inputs and confirmation when devices have some combinations of keyboards and displays. There is also an out-of-band (OOB) mode available for devices that have access to secondary modes of communication, such as the Wi-Fi. A central node can maintain a list of paired and known devices for faster connections. Link communications are encrypted using the 128-bit AES encryption.

RFID

Radio-Frequency Identification (RFID) is a technology for identifying objects using electromagnetic fields that are machine readable. It is an identification mechanism, not a network in a classical sense. However, we mention it briefly because it is widely used and could be a component of IoT systems for asset and object tracking. Various types and modes of uses of RFID are standardized by ISO and ISO/IEC.

In principle, RFID works by tagging objects of interest with RFID tags and using readers to detect them and read their information. Tags contain small microchips that store information, usually an identifier, and an antenna that receives radio transmissions. Passive tags do not contain any power sources. They are powered by harvesting the energy from RFID readers that emit radio waves to detect and read tags.

RFID tags are inexpensive and small, roughly the size of a postage stamp. They can be attached to or embedded in objects to be tracked, such as parts in a production line, livestock, pets, and even people. They have a variety of uses, such as supply-chain management and asset tracking, identification of pets (who have embedded RFID microchips), retail store inventory location and tracking in real time, and many others. RFID tags can act as an electronic equivalent of universal product code (UPC) identifiers, with the added advantage that RFID tags can be read by readers without having to be visible to them (unlike UPC scanners). This is useful in supply-chain management applications where an entire shipping pallet can be tagged and even its individual packages identified without unpacking it as it is entering or leaving a warehouse.

Passive tags are the most commonly used variant of RFID. There are also active tags which are battery-powered to increase range and variants of tags that can be written once or multiple times for identity assignment and modifications after they are manufactured. Being a radio technology, RFID faces a number of practical issues ranging from signal strength and quality to security, as transmissions can be exposed and tags identified by rogue readers.

Long-Range Low-Power Wide-Area Networks

Networks based on IEEE 802.15.4 and their derivatives that we discussed so far are suited for low-power IoT devices in the relative proximity of each other or a gateway. Their effective range is for node to node distances measured in tens of meters, with some extensions possible with mesh configurations. This section discusses long-range low-power wide-area networks operating in unlicensed and licensed portions of the radio spectrum.

Long-range wireless networks are used in applications with geographically dispersed nodes, many of which may be constrained, with distances in the range of kilometers, such as smart cities, agriculture, and open-pit mining. They typically provide lower-bandwidth and higher-latency communications in exchange for the extended range.

LoRa WAN

LoRa (long range) is a low-power WAN specified by the LoRA industry alliance [16]. It operates in the unlicensed spectrum on several bands starting from 169 to 430 MHz, with 868 MHz (Europe) and 915 MHz (North America) being the most common. Its range is on the order of 10 km in open areas and closer to 1 km in densely populated spaces such as cities. The bandwidth is up to 50 Kbps, depending on the distance and transmit power. Payload size is in the range of about 50 bytes to 250 bytes, depending on the frequency band. The network is effectively half-duplex, that is bidirectional communication but in only one direction at a time. Consequently, senders and receivers need to coordinate in taking directional turns.

LoRa specification consists of PHY and MAC layers that can support applications with raw data or IP network layers above. It defines three types of nodes:

- Bidirectional end devices (Class A) – Each device uplink transmission is followed by the two short downlink receive windows. Low-power nodes, no downlink-initiated communication is supported.
- Bidirectional end devices with scheduled receive slots (Class B) – In addition to Class A random receive windows, class B devices open extra receive windows at scheduled times.
- Bidirectional end devices with maximum receive slots (Class C) – Have nearly continuously open receive windows, closed only when transmitting. Usually mains-powered devices, lowest latency for server to device communication.

LoRa uses spread spectrum modulation and supports an Adaptive Data Rate (ADR) algorithm that allows data rates on individual links to vary based on signal strength and radio power. Data rates can range between 0.3 and 27 Kbps. As indicated, Class A nodes can initiate communication with the server when desired with minimal latency, subject only to collisions. However, communication latency from the server can be practically unbounded, as it depends on endpoints to initiate transmissions. Class B nodes rectify this somewhat, but their latency for opening a receive window can be on the order of 128–256 s.

Network topology is star of stars, with end nodes having direct link to radio gateways that use a back-end network to connect to LoRa WAN servers. Being a star with the gateway at the center, nodes can connect with the radio gateway directly and use it as an intermediary to exchange messages with other nodes. Radio gateway operates at the PHY layer only, forwarding messages to the server or other nodes as needed. A gateway is specified to be able to support thousands of nodes.

MAC layer is terminated and processed by the LoRa WAN network server which is connected to the center of the star of stars topology. The network server is required in LoRA installations for MAC layer handling and other management purposes. Network server manages the network including the radio frequency and data rate of each endpoint by means of the ADR algorithm. It also performs frame and address checks and authentication, forwards message to application, and joins servers. Those servers can host applications and manage network joins, respectively.

A single network server can manage multiple LoRA gateways. A LoRA network can and often does include multiple gateways for extended and improved coverage. LoRA nodes do not have a fixed association with a gateway. Consequently, their messages may be received by more than one gateway. One of the functions of the network server is to detect and filter duplicates, check security, and pick the optimal gateway to send acknowledgment.

The LoRa WAN protocol architecture is not based on the Internet network layering. LoRa applications can exchange data with endpoints in raw format with no intervening translations. This, of course, requires intimate knowledge of the endpoint data formats and requires custom coding for each individual type of data source that a destination needs to support. Alternatively, application protocol stacks, such as ZigBee or CoAP, can be layered on the network server or an intermediary and used for data structuring and interoperability with a broader range of devices.

LoRa supports roaming with the use of home server and defined handoff procedures. Devices can be activated by personalization and via over-the-air (OTA) activation that can also deactivate decommissioned or compromised devices. The specification includes definitions of device commissioning and activation procedures.

LoRa networks provide security at the MAC layer that achieves endpoint authentication via the network server. In addition, LoRa WAN packets are protected by symmetric encryption based on AES.

The LoRa alliance provides certification to insure cross-vendor interoperability.

Other IoT LPWANs

There are several other proprietary LPWAN networks providing node interfaces and some parts of the back-end infrastructure for IoT things, such as Sigfox [17] and Ingenu [18]. They are in the process of building infrastructure with access as a service in several countries on the assumption that existence of a low-cost offering for IoT connectivity will foster many IoT installations and lead to the sustainable recurring revenue. This approach seems to be having limited commercial success to date. In general, LPWANs are in competition with telco IoT-targeted offerings described in the next section, and it remains to be seen how they and the market will evolve.

IoT in Licensed Spectrum

As mentioned earlier, use of the licensed portion of the radio spectrum is generally limited to operators that are granted exclusive access to it through government and commercial arrangements. IoT endpoints that intend to use part of the licensed spectrum must make a contractual agreement with its operator to do so. This means that an endpoint needs to contain a suitable hardware interface and modem for the chosen type of the link, as well as be registered by the network operator and provisioned with a SIM (Subscriber Information Module) card or some functionally equivalent mechanism.

The licensed radio spectrum is predominantly used for delivery of voice, data, and video to mobile clients, such as smart phones or tablets. Multiple generations of that technology have been deployed worldwide with ever increasing bandwidth, data rates, and number of users. The 3rd Partnership Project (3GPP), an industry standard development organization for telecommunications [19], has issued numerous system specifications that address radio access networks, services, and core networks. The network was circuit switched for voice and SMS until 3G specifications and then moved to all IP network in subsequent generations with 5G being the most recent one. Some of the goals for the 5G specification include 20 Gbps peak data rate, mobility, i.e., maximum speed for handoff 500 km/h and the density of 10^6/km2 devices. The network is designed to support high-bandwidth dense distribution of clients with high mobility, even when traveling in high-speed trains.

Mobile Cellular Networks

A mobile telecommunication network consists of three basic components (1) endpoint client, (2) radio network, and (3) core network. As shown in Fig. 3.8, the radio network is segmented into individual nonoverlapping cells. Each endpoint client receiving service is connected to the base station of the cell in which it is located at a particular point in time. When the endpoint, i.e., the user with her mobile phone, moves to another area covered by a different cell, the system hands over the subscriber's connection to be serviced and managed by that cell for the duration of residence. Connecting active users and maintaining connections across cells without interruption is the primary management function of a mobile network.

Each radio cell contains a base station with a radio transceiver, set of antennas usually mounted on a tower, and some equipment to manage active endpoint connectivity. The cell designs use methods that minimize interference and cross talk between them, such as using different radio frequencies from the neighboring cells. Those frequencies may be reused in nonadjacent cells. Cell size is limited by the effective range of its transceivers, the topology of the area in which it operates, and the number of subscribers it is supposed to support. Urban areas with high density of subscribers are intentionally divided into cells with reduced range to increase capacity by allowing frequency reuse between nonadjacent cells.

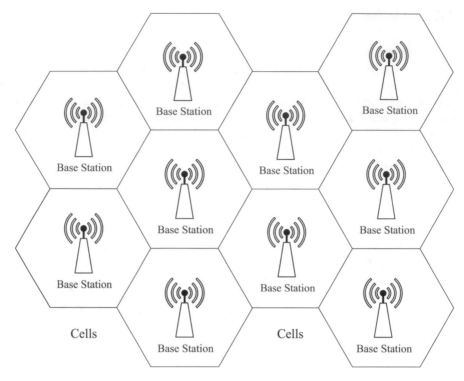

Fig. 3.8 Cellular network

Base stations are powered and connected to the core network that handles complex network signaling and management functions. The core network also handles subscriber management in terms of registration, usage tracking, and billing. On the back end, the core network connects to the public telephone network and to the public data network for global voice and data interchanges. Separate core networks and base stations are usually owned and operated by the telecommunications companies. They may co-locate base stations near the cell towers, sometimes operated or leased by different entities, and have sharing arrangements for handling each other's subscribers while roaming.

With spectrum licenses in place and elaborate infrastructure widely deployed, telco networks are an obvious candidate for handling IoT traffic. With the great expectation of growth in the number of IoT endpoints, they also have a business interest in doing so.

In terms of range with cellular connections, the only important metrics is the supportable wireless distance between the endpoint and the nearest cell tower. Technically, that range can be up to 70 km in areas with no obstacles and clear line of sight, significantly less in hilly and populated areas.

IoT systems can make use of telco networks to provide connectivity to endpoints in geographic locations that are not in the physical proximity of Internet infrastructure. Higher-powered nodes, such as gateways, can use the standard telco data services for connections to the Internet and eventually the cloud. For this reason, many vendors of commercial IoT gateway hardware include model variants with versions of telco modems – such as the 3G, LTE, or 5G – and provisions for SIM cards installation.

Cellular Network Support for Constrained IoT Nodes

The telecommunications industry is also working on options for supporting constrained nodes that may be battery-powered. The intent is to make them more suitable for IoT sensors and things in terms of cost and power requirements of network interfaces and modems while providing adequate bandwidth. Several modifications to endpoints and radio networks are used to address these requirements. They include:

- Reduction of bandwidth, lowering of data rate
- Provisions for sleeping endpoints
- Reduction of radio power
- Use of half-duplex lines
- Payloads restricted to data, no voice or video
- Limited mobility – such as restriction to a single radio cell

Reduction of bandwidth refers to the two techniques that may be used individually or combined. One is reduction of uplink and downlink speeds. Those two speeds may be asymmetrical and differ from each other. The other technique is the use of narrowband channels as opposed to the customary 20 MHz channels. The advantage is that multiple IoT narrowband nodes may share a single wider channel or that narrowband channels can be buried into regular channels or used in the guard band portions of the frequency allocation.

Provisions for sleeping allow endpoints to occupy sleep states for periods of time when they do not participate in paging and network signaling. This requires some redesign since cellular networks usually assume that a node not responding to paging is gone or otherwise unavailable. With IoT, sleeping needs to be supported as a valid state that is distinct and exempted from the default treatment as unreachable and disassociated. This is important because reconnection of a detached mobile node is a somewhat lengthy control and signaling process that increases latency and consumes power.

Reduction of radio power requirements for endpoint transceivers is one of the techniques to conserve energy. Since IoT nodes operate in narrower bands with lower data rates, lower radio power may not significantly impact the range.

Use of half-duplex lines can simplify modems and interfaces, thus reducing costs. It still allows bidirectional communication, although in only one direction at a time.

Limited mobility restriction usually means that an endpoint is confined to using a single radio cell. This works well for nodes that are static or can move only in a limited area within the range of its associated radio cell. This simplifies node management and system signaling by eliminating the need to support handoffs.

Several variations and generations of 3GPP specifications have tackled perceived constrained node IoT requirements by using combinations of these techniques. LTE Cat (category) 0 node specification is for the use of reduced speed of 1 Mbps in half-duplex mode in a 20 MHz channel and with some provisions for idle mode and sleep time. LTE-M category lowers the channel bandwidth to 1.4 MHz and reduces the data rate to 200 kbps in half-duplex mode and with the sleeping time between paging cycles increased to minutes versus seconds for Cat 0. Narrowband IoT or NB-IoT uses the even lower channel bandwidth of 180 kHz at half duplex with data rates of 30 Kbps for downlinks and 60 Kbps for uplinks.

This area is changing, and new specifications are being devised with the evolving understanding of IoT requirements and of business potential for mobile and telecommunications networks in supporting constrained nodes. From a user's point of view, the costs of modems and interfaces are still comparatively high, and the total cost of ownership includes recurring service payments for endpoints that is not required for networks that use unlicensed spectrum. On the other hand, use of the licensed spectrum from a commercial service provider can have a number of benefits. They include potentially broad geographic coverage for mobile endpoints, such as transportation vehicles, bandwidth guarantees with no collisions, and a communications network professionally managed by a third party.

Practical problems include the fact that 3GPP IoT specifications are not yet widely implemented by the operators, and their networks, including cells and base stations, may need to be upgraded to support the desired IoT modalities. Doing so can be costly and usually requires the promise of a larger installed base to justify the business expense to the operator. The opposite side of the problem is that the installed base cannot develop until the service is available.

Other problems that need to be addressed for wider IoT deployments include the need to adapt telco billing models to usage modes and cost tolerances of IoT with large number of endpoints with relatively low data volumes. A somewhat related problem is the management of SIM cards. Installing and provisioning of SIM cards to a large population of nodes is labor-intensive, error-prone, slow, and expensive. A better alternative is to use the electronic SIM cards, or eSIMs. They can be provisioned remotely and in an automated manner. However, their adoption and operator support are still early in the adoption cycle and are not widely available.

Constrained Application Protocol (CoAP)

HTTP is one of the most widely used application protocols on the Internet and the World Wide Web. It is commonly deployed for transfers of HTML and XML documents between endpoints in a client-server fashion. It works well for fully capable nodes, but it can be too demanding in terms of bandwidth and processing requirements for constrained IoT devices. Constrained Application Protocol (CoAP) was developed [20] to address the needs of constrained devices and networks. It uses smaller and simplified headers and supports asynchronous REST and publish/subscribe message exchanges in direct machine-to-machine (M2M) communication. It also supports built-in discovery of services and resources and support for URIs and web media types. CoAP is intentionally designed to be a lightweight counterpart of HTTP and to stay as close as possible to its design in order to simplify integration with the web, while meeting additional IoT requirements such as multicast support, low overhead, and simplicity.

CoAP Messaging

CoAP messaging is designed to suit the needs of M2M and IoT exchanges of messages that primarily consist of sensor data readings and commands. Logically, CoAP has two layers – an upper layer for requests and responses and a lower messaging layer – both of which reside above the UDP transport. It uses UDP at the transport layer between endpoints for simplicity and lower latency, at the expense of reliability guarantees. To cope with the latter, CoAP includes an option for reliable delivery of a class of its messages by means of explicit endpoint acknowledgments. CoAP defines six types of messages as listed in (Table 3.3).

Messages exchanged via CoAP are of the request-response type. Requests are sent by clients to request an action using a method code (explained later in this section) on a resource identified by URI. The server sends a response that may include a resource representation, such as an observed state. Unlike HTTP, these messages are sent asynchronously. Messages contain IDs to detect duplicates and for acknowledgment matching in reliable delivery. They also contain separate tokens to match requests with responses.

The message flow depicted in Fig. 3.9 illustrates an asynchronous message sequence with a separate response. The GET request is sent as a confirmable message. However, the server cannot provide an immediate response, say due to

Table 3.3 CoAP message types

Message type	Description
Confirmable	Reliable message delivery, recipient required to confirm by sending acknowledgment
Non-confirmable	Not acknowledged, delivery not guaranteed
Acknowledgment	Confirms delivery of a specific confirmable message
Reset	Indicates a message was received, but recipient is missing some context to process it properly
Piggy-backed response	Included in the acknowledgment message, provides data response, e.g., sensor reading
Separate response	Sent as a delayed response to a prior request that needed time to process

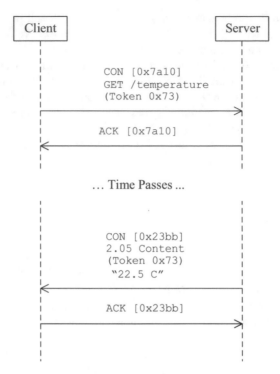

Fig. 3.9 CoAP asynchronous message exchange with a separate response

unavailability of a fresh temperature reading, and it just sends an acknowledgment to confirm the receipt. Some time later when the data become available, it is sent by the server in a separate confirmable message with success indicator and the required content. That message is correlated to the original request by carrying the same token, 0x73. Note that message identifiers are used just to match request messages with immediate responses, such as the confirmation of request receipt, ID of 0x7a10 in this example. The token, on the other hand, is used to correlate the asynchronous response to the original request, so it is the same in GET and content response messages, in this example with the value of 0x73.

In further alignment with the mainstream web, CoAP uses standard URIs (Uniform Resource Identifiers) to identify and address target entities in requests.

CoAP uses well-defined methods with which clients can invoke actions on the target resource, identified by URI, at the server to which the request is sent. CoAP methods are intentionally similar to HTTP as described in Table 3.4.

These methods conceptually correspond to those used by HTTP. The difference between POST and PUT is somewhat subtle – PUT is generally used to replace the existing content of a resource, while POST is used to send new data to a general point in the hierarchy where a place for it may need to be created. The confusing part may be that both can result in the creation of a new target resource if it does not exist. The difference is that PUT identifies a specific URI to be updated or created, whereas POST URI specifies a resource that should deal with the request, possibly

Table 3.4 CoAP methods

Method	Description
GET	Retrieves a representation of information corresponding to the specified URI
PUT	Requests that the identified resource be updated with the enclosed representation, optionally created if it does not exist
POST	Request that the enclosed representation be processed, e.g., new resource created or existing updated
DELETE	Deleted resource identified by URI

creating a new entry or level of hierarchy to store the data. Consequently, PUT operations is idempotent meaning that its multiple requests with the same payload to the same target will produce the same result. POST operation is not idempotent in the sense that its repeated multiple requests will have the side effect of creating the same resource multiple times. Some IoT frameworks favor the use of CRUD methods (CREATE, RETRIEVE, UPDATE, DELETE) that come from the storage systems and avoid this ambiguity.

CoAP supports broadcast and multicast to deliver messages to multiple nodes in a single transmission. One of their uses is for discovery where a requesting node solicits description of access points and services from other nodes. The CoAP specification also supports proxying and caching. In general, proxies are intermediaries that can perform translation of protocols and namespaces and optionally cache the content. Cashing refers to the temporary storage of content at intermediate points that can satisfy closely spaced requests faster without having to retrieve the content all the way from the original source. Cached content is usually accompanied by a freshness parameter that indicates its expiration time after which it needs to be discarded. CoAP specification also describes cross-proxies that can perform translations between endpoints that use different application protocols, such as CoAP to HTTP and vice versa.

In terms of security, CoAP communications can be secured at the transport layer, such as DTLS for UDP, as well as at the link layer as described earlier in conjunction with IEEE 802.5.14. Additional IETF specifications describe the use of CoAP over TCP and with web sockets [21].

Messaging and Queuing

Application protocols such as CoAP and HTTP are primarily designed for request-response type of exchanges between clients and servers. In typical IoT uses, servers own resources, and clients issue requests to retrieve or update their state. In most cases, this is a one-to-one communication initiated by clients. To retrieve data, clients need to request them, thus effectively dictating the pace of data collection. This is sometimes referred to as the *pull* mode of communication, where clients pull the data of interest when needed.

An alternative mode of data reporting, sometimes referred to as the *push* mode, is for the server to report data when sampled or changed in a pre-defined manner.

For this to work, servers need to keep track of clients with known interest in a particular data object or attribute. This can be a demanding task for constrained nodes. A publish-subscribe mechanism, described below, may be used to alleviate the problem.

A data exchange mechanism that became quite popular on the Internet is the publish-subscribe pattern where data sources, called publishers, post data at a predefined aggregation point and interested receivers, called subscribers, can subscribe to receive data of interest. The mediating function is usually delegated to intermediate nodes, called brokers, that receive published messages and deliver them to registered subscribers. Publish-subscribe mechanism has some characteristics that can be useful in IoT systems including:

- Decouples publishers from subscribers
- Allows for asynchronous communication
- Facilitates scalability

Publishers are decoupled from subscribers in the sense that they only need to publish messages to a known point, such as a broker port, and do not need to know the identities or the number of their subscribers, if any. Subscribers are decoupled from publishers in the sense that they do not need to know the identity of their data sources nor engage with them in the explicit request-response exchanges. One of the advantages of this arrangement is that subscribers may be added to or removed from the system in the course of its lifetime without impacting the code or behavior of data sources.

Communication is asynchronous in the sense that concurrent availability of clients and servers is not required to complete the data exchange. This requires brokers to buffer messages until they are received or, for a period of time, whatever may be specified by the selected quality of service and system policies in effect.

The basic operation and components of a brokered publish-subscribe system are illustrated in Fig. 3.10. It shows message queues maintained by the broker, publishers, and multiple subscribers to topics. Publish-subscribe systems usually include mechanisms to filter messages thus allowing subscribers to receive only messages of interest. This is accomplished by publishing messages to named logical channels, called *topics*. A topic can be a point of interest, such a specific sensor, or a group of sensors attached to a gateway or belonging to a domain. Subscribers specify topics that are of interest to them and receive only the related messages, such as on topic x or topic y in Fig. 3.10. A subscriber can subscribe to multiple topics. A publisher publishes messages to a topic, and they are delivered to all subscribers of that topic. Brokers usually handle matching and filtering of topics.

In IoT systems with constrained nodes, this allows a data source to publish a message to a potentially large number of subscribers via a single message exchange with a broker. The result is reduced load on the data sources (IoT servers) and potentially increased scalability.

Message brokers maintain connectivity with publishers and subscribers, keep track of active subscribers, filter messages and topics, and store messages as

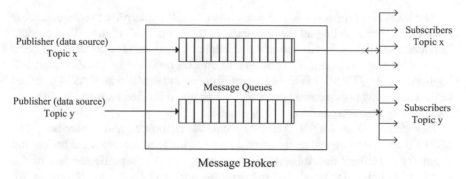

Fig. 3.10 A brokered publish-subscribe system

necessary for confirmed delivery when required. They are usually installed in power-affluent nodes with comparatively good network connectivity, such as the high-end gateways, fog nodes, and the cloud.

One practical example of the value of publish-subscribe and decoupling in particular is that it makes it possible to add major new functionality to a running system without disrupting its operation. For example, an entire sensor database may be added to a system that does not have it by installing the database at some point and having it subscribe to all relevant topics. In practice, this would mean creating some type of a writer service that subscribes to all sensor and event topics of interest and writes the received data into the database. Moreover, the database can be subsequently replaced by another one or even completely removed without the awareness or disruption of publishers. An example of how this can be implemented is provided in the case study in chapter "Putting It All Together".

There are several messaging and queuing systems with some degree of suitability for use in IoT systems, such as AMQP [22], DDS [23], and XMPP [24]. In the next section, we describe the MQTT as representative of this approach because it was specifically designed for IoT systems, and it is widely used.

Message Queuing and Telemetry System (MQTT)

MQTT (Message Queuing and Telemetry System) is a publish-subscribe system specifically designed for IoT systems including those with constrained nodes. It is widely used in IoT platforms for delivering edge sensor data to applications and services. It is the default data ingestion mechanism used by the commercial cloud platforms described in chapter "IoT Platforms".

MQTT is designed for use in networks that may have low bandwidth, high latency, and fragile connections among constrained devices with small memory and low power. In practice, it has been used in heterogeneous populations of publishers and subscribers that include constrained things, fully fledged IoT gateways, fog nodes, and cloud data ingestion points.

The MQ part of the name is a remnant from the IBM enterprise message queuing system MQ Series dating all the way back to the era of mainframes. That product provided the original inspiration and operational experience for the definition of a lightweight version aimed for sensors and constrained devices in M2M and IoT applications. MQTT 3.1.1 is a public specification available as an OASIS standard [25]. It has many open-source implementations, and it is often included in commercial gateway implementations.

Key players in an MQTT system are publishers, broker, and subscribers. The MQTT documentation and many descriptions use the term server for the broker and client for publishers and subscribers, depending on which particular end of the exchange is being described, i.e., publisher to broker or broker to subscriber. We will use the term publisher for message sources (IoT server) and subscriber (IoT client) to maintain consistency with the rest of the text and to avoid confusion.

MQTT assumes that the network connections among its components provide the means to send ordered, lossless streams of bytes in both directions. In practice, many implementations of MQTT use TCP to meet this requirement. The protocol is lightweight in the sense of requiring only a 2-byte header for messages, with some optional extensions. It defines 14 message types for control and for publishers and subscribers to use. Some of those messages may contain payloads for (broker relayed) exchanges between publishers and subscribers. MQTT is protocol agnostic in the sense that it forwards but does not interpret message payloads. Consequently, the payload can be a binary encoding or some form of human-readable serialization, basically anything that can be produced by publishers and understood by subscribers.

The MQTT broker keeps track of the states of publishers and subscribers and handles some cases of unreachability or disconnectedness. For this purpose, it includes ping-like control messages to check liveness of connections that are inactive beyond configurable keep-alive intervals.

MQTT provides a choice of three different levels of quality of service for message delivery as follows:

- QoS 0: at most once delivery – best effort message delivery with no confirmation of receipt.
- QoS 1: at least once delivery – message receipt is acknowledged, duplicates are possible.
- QoS 2: exactly once delivery – message receipt confirmed, no duplicates.

The overhead increases with the increasing QoS levels. QoS 1 requires an acknowledgment message and QoS 2 requires a multi-step acknowledgment between a publisher and a broker and then on with the subscriber. As a simplification, the broker can acknowledge a successful message receipt to the publisher and then assume the ownership of the message until its delivery is completed.

MQTT uses 14 types of messages, designated by the numeric value provided in the header. As indicated in Table 3.5, they provide the means to establish and break connections, publish and subscribe to messages, various forms of acknowledgment, and ping messages to check liveness of connections. The full complement of

Table 3.5 MQTT messages

Message type	Value	Description
CONNECT	1	Request to connect
CONNACK	2	Connect acknowledgment
PUBLISH	3	Publish message
PUBACK	4	Publish acknowledgment
PUBREC	5	Publish received
PUBREL	6	Publish release
PUBCOMP	7	Publish complete
SUBSCRIBE	8	Subscribe request
SUBACK	9	Subscribe acknowledgment
UNSUBSCRIBE	10	Unsubscribe request
UNSUBACK	11	Unsubscribe acknowledgment
PINGREQ	12	Ping request
PINGRESP	13	Ping response
DISCONNECT	14	Disconnect

message processing is usually supported by brokers, but implementation of constrained clients can be simplified by omitting some that they do not need to support. For example, simple applications that need only the lowest quality of service can be implemented by clients that implement and use just the CONNECT, PUBLISH (for publishers), SUBSCRIBE (for subscribers) and DISCONNECT messages.

MQTT also supports the use of the "will" message (as in the last will) that can be associated with a link and be published to subscribers to indicate the last action or notification to deliver when the link is closed. The link can be closed via a control message or due to errors, such as failing to send a ping request before expiration of its time-to-live interval.

A broker (MQTT server) accepts network connections from clients, accepts application messages that are published, manages subscription requests, and forwards application messages that match client subscriptions. It also keeps track of topic names and filters them as necessary. The broker stores message in transit until they are sent to all subscribers, and the acknowledgment sequences dictated by the QoS level in effect are completed, if any. Moreover, when instructed by a flag in the header, the broker retains a message and sends it to new subscribers immediately so that they receive a replay of the most recent message on the topic and do not need to wait until the next "fresh" publication on that topic arrives.

Topics are filtered by name, and they can have levels of hierarchy denoted by "/." Wild cards are supported to allow subscriptions to multiple topics. For example, a subscriber to the topic

```
building5/floor3/room305/temp/t1
```

will receive all messages published to it, i.e., temperature readings from the sensor t1.

A wild card "#" matches any number of levels within a topic. It allows subscriptions to a range of topics. For example, a subscription to

```
building5/floor3/room305/temp/#
```

will get all qualifying readings at lower levels in the hierarchy, e.g., all temperature readings from the room 305, including t1 and all levels below it, if any. Similarly, subscription to

```
building5/floor3/#
```

will get all sensor readings from the third floor in the building five. Another supported wild card is the plus sign "+" that allows matching of only one level of the topic.

Use of MQTT requires deployment of broker nodes that buffer messages and keep track of topics and active publishers and subscribers. Brokers may reside on any sufficiently resourced node, including gateways, fog nodes, and the cloud. Complexity and size of the broker implementation depends on the chosen operational options. For example, exactly once semantics and longer retention of messages can increase bandwidth, complexity, and consume more resources. Most of that burden is borne by the MQTT broker nodes, not by publishers and subscribers.

References

1. Tanenbaum, A., Wetherall, D. (2011) *Computer Networks*. 5th ed., Boston, MA: Prentice Hall.
2. Day, J., Zimmermann, H. (1983) 'The OSI reference model' *Proceedings of the IEEE*, 71(12), p 1334–1340.
3. The Industrial Internet of Things Vol G5: Connectivity Framework (2017) [Online] Available at: https://www.iiconsortium.org/pdf/IIC_PUB_G5_V1.01_PB_20180228.pdf (Accessed Dec 15, 2019)
4. Borman, C., Ersue, M., Keranen, A. (2014) 'Termino0logy for constrained networks', IETF RFC 7228. [Online] Available at: https://tools.ietf.org/pdf/rfc7228.pdf (Accessed Dec 15, 2019)
5. IEEE 802.15.4-2003 – IEEE Standard for Telecommunications and Information Exchange Between Systems – LAN/MAN Specific Requirements – Part 15: Wireless Medium Access Control (MAC) and Physical Layer (PHY) Specifications for Low Rate Wireless Personal Area Networks (WPAN) [Online] Available at: https://standards.ieee.org/standard/802_15_4-2003.html (Accessed Dec 15, 2019)
6. Montenegro, G. et al. (2007) 'Transmission of IPv6 packets over IEEE 802.15.4 Networks', IETF RFC 4944. [Online] Available at: https://tools.ietf.org/pdf/rfc4944.pdf (Accessed Dec 15, 2019)
7. Culler, D. E., Hui, J., Shelby, Z. (2010) 6LowPAN [Online] Available at: https://people.eecs.berkeley.edu/~culler/talks/IPSO-6LoWPAN.ppt (Accessed Dec 15, 2019)
8. Winter, T. et al. (2012) 'IPv6 routing protocol for low power and lossy networks', IETF RFC RFC 6550 [Online] Available at: https://tools.ietf.org/pdf/rfc6550.pdf (Accessed Dec 15, 2019)

9. Thread group [Online] Available at https://www.threadgroup.org/ (Accessed Dec 15, 2019)
10. HART communication protocol specifications [Online] Available at: https://fieldcommgroup.org/hart-specifications (Accessed Dec 15, 2019)
11. ISA 100 specification [Online] Available at: https://www.isa.org/isa100/ (Accessed Dec 15, 2019)
12. Zigbee specifications [Online] Available at: https://zigbee.org/zigbee-for-developers/zigbee-3-0/ (Accessed Dec 15, 2019)
13. Perkins, C., Belding-Royer, E., Das, S. (2003) 'Ad hoc on-demand distance vector (AODV) routing', IETF RFC 3561, [Online] Available at: https://www.rfc-editor.org/rfc/pdfrfc/rfc3561.txt.pdf (Accessed Dec 15, 2019)
14. Bluetooth specifications [Online] Available at: https://www.bluetooth.com/specifications/ (Accessed Dec 15, 2019)
15. Nieminen, J. et al. (2015) 'IPv6 over Bluetooth low energy', IETF RFC 7668 [Online] Available at: https://tools.ietf.org/pdf/rfc7668.pdf (Accessed Dec 15, 2019)
16. LoRA WAN [Online] Available at: https://lora-alliance.org/sites/default/files/2018-04/what-is-lorawan.pdf (Accessed Dec 15, 2019)
17. Sigfox LPWAN [Online] Available at: https://www.sigfox.com/en (Accessed Dec 15, 2019)
18. Ingenu LPWAN [Online] Available at: https://www.ingenu.com/ (Accessed Dec 15, 2019)
19. Third Generation Partnership Project, 3GPP [Online] Available at: https://www.3gpp.org (Accessed Aug 15, 2019)
20. Shelby, Z, Hartke, K., Borman, C. (2014) 'The constrained application protocol (CoAP)', IETF RFC 7252 [Online] Available at: https://tools.ietf.org/pdf/rfc7252.pdf (Accessed Dec 15, 2019)
21. Borman, C. et al, (2018) 'CoAP over TCP, TLS and websockets', IETF RFC 8323 [Online] Available at: https://tools.ietf.org/pdf/rfc8323.pdf (Accessed Dec 15, 2019)
22. AMQP [Online] Available at: https://www.amqp.org/ (Accessed Dec 15, 2019)
23. DDS [Online] Available at: https://www.omg.org/spec/DDS/1.4/PDF (Accessed Dec 15, 2019)
24. XMPP [Online] Available at: https://xmpp.org/ (Accessed Dec 15, 2019)
25. OASIS Standard, MQTT Version 3.1.1 (2014) [Online] Available at: http://docs.oasis-open.org/mqtt/mqtt/v3.1.1/os/mqtt-v3.1.1-os.pdf (Accessed Dec 15, 2019)

Chapter 4
Cloud

Cloud is the point of high-level aggregation and processing of data in an IoT system. As indicated earlier, data aggregation and processing may take place at multiple levels in the IoT system hierarchy, but lower levels operate with limited data sets. Depending on the implementation, a cloud-level aggregation can span multiple IoT domains to provide system-level actionable insights within an enterprise or segments of an industry.

Functionally, the cloud is the meeting place of large-scale IoT data and applications and services that operate on them. Streaming and archived data are made available to the authorized users via APIs and queries. Cloud implementations generally host the top-end data storage and access methods and provide the server, storage, and connectivity infrastructure to execute services and applications that use them.

In this chapter we review cloud technology and implementations of the cloud portions of IoT systems, including data ingestion, storage, and processing. The second part of the chapter is an overview of the machine learning and artificial intelligence techniques that provide much of the foundation for analytics that creates insights and recommends or automates the resulting actions in IoT systems.

Cloud Computing

Cloud computing is a model and technology for enabling on-demand access to configurable computing resources including servers, storage, network, and services [1]. Cloud hardware resources are virtualized and managed in a way that automates allocation of resources and quickly responds to resource allocation requests, sometimes on the order of milliseconds. Cloud servers and storage are concentrated in data centers, and they can be allocated to services and applications in a dynamic and elastic manner. Elasticity means that the amount of resources allocated to a service can grow and shrink dynamically to adjust to the changes in load and demand.

© Springer Nature Switzerland AG 2020
M. Milenkovic, *Internet of Things: Concepts and System Design*,
https://doi.org/10.1007/978-3-030-41346-0_4

Cloud systems provide externally visible access points with ample bandwidth for data ingestion. Additional access points are available for uploading and activating user applications and services in addition to those that may already be deployed by the cloud provider or by other commercial sources. Users may access services via the Internet to configure, run, monitor, and manage execution of their applications.

Cloud services provided and hosted by the commercial cloud providers are often referred to as *public clouds*. They are public in the sense that almost anyone with a credit card can access and use one for a fee. The same technology may be used to manage computing resources within an enterprise. Such installations are usually called *private clouds*. They provide greater degree of control and potentially privacy and security but tend to have comparatively smaller resource pools which may reduce scalability and elasticity. The term *hybrid cloud* refers to installations that use the same technology as the chosen commercial provider and divide their data and resources between the public and the private portions to meet security, regulatory, and performance demands.

From a user's perspective, key features and benefits of cloud computing include:

- Appearance of availability of infinite computing resources on demand
- Scaling up and down in accordance with load variations
- No up-front capital expenses for computing hardware
- Simplified operation due to virtualization and professional management
- Potentially increased reliability, resilience, and security

Most of those benefits result from the use of virtualization and multiplexing of workloads across a potentially large number of users. Virtualization decouples the workload from the physical server by encapsulating it in a virtual machine (VM) or containers. Cloud tools enable rapid automated provisioning and activation of VMs on servers in the pool. This enables cloud-based applications and services to scale up quickly to respond to spikes in demand, giving the illusion of infinite resource availability. As an additional benefit, applications can scale down responsively when the load subsides in order to reduce the operational costs. Cloud charges are typically based on the actual, accounted for, use of resources, including VMs, storage, and bandwidth.

Running of multiple VMs and containers on a single physical server can increase its utilization and provide a better return on hardware investment. Cloud providers can multiplex workloads from a large number of customers that tend to have less variance in the aggregate. These efficiencies and economies of scale allow cloud operators to achieve higher resource utilization and be more cost-effective in comparison with the traditional enterprise IT systems.

One of the appealing aspects of cloud computing is the ability to pay for the use of resources on the as-needed ongoing basis without the up-front capital expense for purchasing the hardware. This can be especially attractive for start-ups and smaller companies as they do not have to go into debt for capital expenses and can pay for the use of cloud resources from their ongoing revenue. Moreover, they are relieved of the burden of accurately estimating their needs for fixed computing capacity. If the capacity estimate exceeds the actual need, money will be spent on the unused equipment, and the return on capital investment will be low. If the need is underestimated, the company will not be able to meet the demand, resulting in the loss of revenue and unhappy customers who may choose to go elsewhere. For these reasons, venture capitalists in Silicon Valley have all but banned the line item for server purchases in start-up business plans years ago.

The high concentration of computing resources in cloud data centers, coupled with virtualization and automated management tools, makes it possible to provide quality professional system and security management often at lower cost in comparison with the on-premises enterprise IT installations.

Major cloud providers tend to have data centers in multiple geographies. This can help with the regulatory compliance for data that need to be kept in particular legal jurisdictions. Other potential benefits include reduction of network latencies and costs by judicial placement of data centers and disaster protection by replicating critical data and services across regions.

Downsides of cloud computing include loss of function when there are network or cloud outages and potentially increased concerns for security and privacy protection of sensitive data that is kept off the user premises. Their usage incurs low upfront but recurring and variable costs that need to be weighed against other infrastructure ownership and operation options. Another potential problem is the vendor lock-in. Cloud systems and offerings are mostly proprietary, and changing vendors to reduce cost or to get better service can be very difficult and expensive in terms of porting applications and migrating data to the new provider.

Models of Cloud Computing

Cloud installations and offerings provide several different levels of service abstractions and models that are illustrated in Fig. 4.1. The original model, often referred to as the *Infrastructure as a Service* (IaaS), basically works with a cloud vendor

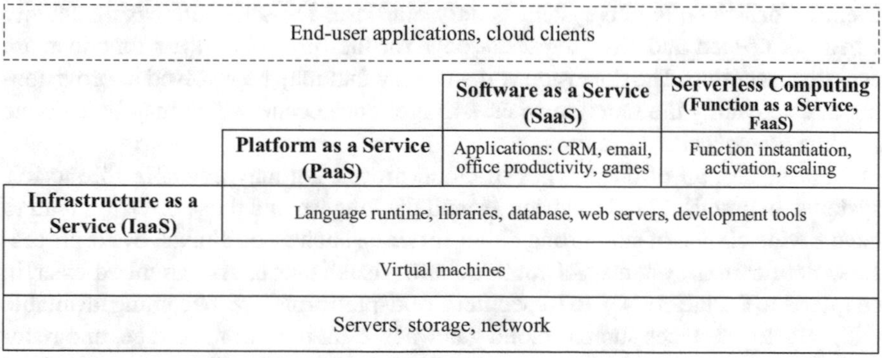

End-user applications, cloud clients		
	Software as a Service (SaaS)	**Serverless Computing** (Function as a Service, FaaS)
Platform as a Service (PaaS)	Applications: CRM, email, office productivity, games	Function instantiation, activation, scaling
Infrastructure as a Service (IaaS)	Language runtime, libraries, database, web servers, development tools	
Virtual machines		
Servers, storage, network		

Fig. 4.1 Types of cloud offerings

providing managed infrastructure in terms of VMs for rent, networking, and some forms of block and object storage that users can request. In this model, users are responsible for providing VM images for execution that may include an operating system, runtime environment, and applications. Consequently, users are responsible for patching software and monitoring their VM's operational status.

A higher level of abstraction is provided by the *Platform as a Service* (PaaS) where users provide their own applications created using programing languages, services, and tools supported by the provider. In this model, the platform provider manages the underlying infrastructure as well as the functioning and updates of the runtime system and supporting services, thus relieving the users of that burden.

Software as a Service (SaaS) refers to systems where an entire application, such as the office productivity tools or a customer relationship management (CRM) system, is available from the provider and customers can connect to it using clients such as web browsers and mobile applications.

A variant called *Back end as a Service* (BaaS) originated with mobile systems to enable application developers to focus on the device and front-end part of the application, while the service provider manages the cloud side, including storage and features such as user authentication, push notification, and interface to social networks or location services.

Function as a Service (FaaS) is a service model in which the user provides a function, i.e., a piece of code in a supported programming language that needs to be executed in response to a defined triggering event. For example, the function may provide image classification and be triggered when an image is uploaded to a database. In this model, the service provider is responsible for creating an execution environment, provisioning, and activating the function when its associated triggering event occurs. The provider is also responsible for maintaining scalability and resilience by activating and managing multiple instances of the function as necessary. A combination of the FaaS and BaaS is referred to as *serverless computing* [2]. In this model, the user defines the processing function, and the service provider manages everything else – its instantiation, scaling up and down, and the entire underlying infrastructure. One of the characteristics of serverless computing is that no active instances of functions are maintained in the absence of invoking events. Consequently, the system is very elastic in the sense of ceasing activity when not needed and charging users only for the time when their functions are actually executing. The downside is the latency that may be involved in provisioning and activating the function when a trigger does occur, which may be an issue in some time-sensitive IoT uses.

In principle, all of these types of cloud processing may be applicable to IoT systems. In practice, IaaS has been around the longest, and those offerings tend to have a wide choice of supporting cloud services, such as databases, event processors, notification systems, AI toolkits, and visualization. As described later in chapter "IoT Platforms", IoT-specific cloud platforms are becoming available with targeted services such as cloud gateways, data ingestion, storage, and security. There are some nascent offerings of IoT PaaS and SaaS that are somewhat experimental in terms of features and business models. Serverless computing is supported by the commercial providers both in the cloud and at the edge. This allows flexible placement of functions, thus facilitating implementations of the edge-to-cloud computing continuum in IoT systems.

IoT System Cloud Components

Implementations of core support for IoT systems in the cloud provide the infrastructure and services for acquiring, preprocessing, routing, and storing data for subsequent processing by the cloud applications and services. They typically include the following components:

- Edge interface and data ingestion
- Time-series stream processing
- Data aggregation and storage
- Security and management

Major functional blocks that commonly implement these functions are depicted in Fig. 4.2.

The box labeled cloud edge gateway is a functional module hosted in the cloud that interfaces with edge devices and things. It also acts as a security boundary and a control point between systems at the edge and the inner core components of the cloud.

Stream data processing and storage are shown as the two parallel paths that may supply data to the back-end applications and services. Commensurate with their processing needs, ingested data may go through either or both paths, to be processed as an incoming stream, stored for archival and batch processing purposes, or both. These two paths are sometimes referred to as the *warm path*, for streaming data, and the *cold path* for storage and the more time-consuming and comprehensive types of processing, such as the analytics and ML model creation described

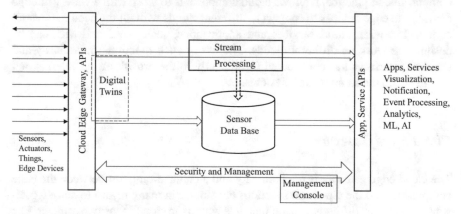

Fig. 4.2 IoT cloud components

later. This dual-path design is sometimes referred to as the *lambda architecture*, ostensibly because its split stream/storage data path resembles the Greek letter lambda, λ, (turned sideways).

Some IoT cloud systems implement digital twins as cyber representations and cloud replicas of endpoints. Their use and characteristics are described in the next section.

Security and management is a crosscutting system function that implements control plane and can touch most of the other functional blocks. It is in charge of keeping the IoT system itself and its components operational and secure. As discussed in chapter "Security and Management", this subsystem tends to have its own agents that are installed on the monitored components to report the state, to perform updates, and to maintain and enforce system security policies.

Figure 4.2 also contains block labeled apps, Services APIs that represent cloud interfaces for the back-end applications and services to access and retrieve IoT data. It may also represent a security boundary and access control point for service authentication and authorization. This is a conceptual representation, and in many contemporary implementations, cloud service interfaces tend to be bespoke APIs provided by the individual applications, such as stream event processors and specific databases.

However, it is a good design practice to define more formalized interfaces and functional API definitions that are not application specific. Retrieved data may be obtained from the real-time streams or storage. The salient point is to formalize the interfaces and service access points and methods to retrieve data. The data themselves may originate from the incoming streams or from the archival storage. Stored data may be in a data lake or databases co-located with the access point or may be distributed and retrieved from various other aggregation points in the system, such as edge of fog nodes.

The same is true for the data ingestion side that should be designed using the more generalized interfaces to edge devices, such as smart things and gateways. This approach allows for the creation of portable applications and interchangeable components. In practice, it allows cloud components to work with a variety of edge devices and edge devices to connect to different clouds without the need for custom coding. Likewise, cloud services and applications, such as AI and ML, can be designed to work on different clouds. The use of such functional modularity and standard interfaces has contributed significantly to the growth of the Internet and will likely do the same for IoT systems.

Cloud Edge Gateway

The cloud edge gateway is the boundary and a delineation point between the edge components and the cloud. Its functions are similar in many regard to an IoT gateway, with the key difference that it operates within the cloud security perimeter. The cloud-to-edge gateway or hub is the gating point between the external edge components with varying levels of security capabilities and the highly secure and trusted cloud execution environment. It typically implements security checks such as access authorization, authentication, and possibly inspection and filtering of inbound data flows and messages based on their senders, content, and intended operations.

Functionally, the IoT hub is tasked with acquiring data and issuing commands to the endpoints connected to it. It may perform these functions for relatively large numbers of endpoints, including sensors, smart things, and edge gateways with their associated sensors and actuators. The IoT hub provides the necessary data access points and ingestion mechanisms, and it might also implement digital twins. Those functions are described in sections that follow.

IoT Data Ingestion

Edge connectivity provides access points for receiving data from the edge and for relaying actions to the edge components, such as the actuation commands and configuration of operational set points. Data arriving from the edge may be coming directly from the individual sensors and things or from IoT gateways that may perform some level of preprocessing and aggregation. Incoming data are either readings of sensor data sampled at regular intervals or asynchronous (irregular) notifications that may include monitored values that exceed thresholds or some composites that constitute events. Input payloads typically contain time stamps and metadata that may further describe the origin and context in which the data was acquired.

Primary functions of cloud data ingestion are to provide access points and methods for the edge sources to deliver data to and to provide the scaling necessary to handle the offered data volumes.

Ingestion of input data primarily includes dealing with volume and routing to the appropriate destination. In terms of the basic connectivity, a cloud interface needs to provide access points and methods suitable for the nature of incoming data payloads. In general, each data source reporting to the cloud is provisioned with a cloud access point for its mode of reporting.

Access points may be implemented as URLs for HTTP posts, web sockets, or message brokers for data streams and messages. They are commonly IP addresses or Internet-resolvable resource identifiers with data ports corresponding to the transport protocols in use.

The simplest form of data ingestion is *direct posting* where data sources, such as sensors or things, post their messages to a designated cloud entry point. This may be an address of a web server that accepts messages from the authenticated endpoints. Depending on the implementation, such entry points may be directly connected to the cloud services that process data from those particular sources or to the aggregation points for multiple services. In the latter case, incoming messages have to be routed to the appropriate services, based on the message type, topic, or content. Ingested data are typically directed to the next data processing stage in the path, such as the stream processing pipeline in the warm path or the database storage writer in the cold path.

IoT cloud edge services may also include hosting of message brokers for the publish-subscribe mode of data delivery. This includes implementation of protocols such as MQTT, management of subscribers, and storage of messages until delivery or as determined by the chosen quality of service.

Data format translation may need to be performed for some messages if there are differences in formats used by the origin of data and the processing service. This function needs to be performed at an optimal point on the data path – at the IoT gateway, at the data ingestion stage, or in the receiving application itself. These considerations and implementation mechanisms are described in the section on interoperability in the chapter "IoT Data Models and Metadata" on IoT data models.

The data volumes that cloud access points need to handle may be very high. Even when the individual sources have a relatively low frequency of reporting, aggregate data rates from a large number of endpoints can be considerable. In order to support high data ingestion rates, cloud access services may need to implement some form of scalability techniques, such as the creation of multiple instances of input servers or message brokers and balancing the load across them. A number of load balances and IP sprayers are available from commercial providers and as open-source implementations. There are also specialized systems, such as Kafka, that can provide scalable data ingestion. Kafka [3] is a high-throughput implementation of a durable messaging and publish-subscribe system that provides high reliability and scalability through partitioning and replication. It was originally designed for handling high-volume web streams, such as social network postings.

Digital Twins

The term digital twin generally refers to a synchronized cyber representation and replica that mirror the states and behaviors of a physical thing. It was originally coined by NASA to describe the hybrid replicas of their vehicles that combined physical artifacts, precise multi-physics models, and live sensor data to mimic its physical counterpart. It was used to simulate and predict behaviors of things in remote locations, such as celestial bodies. For example, a digital twin in the control center could be used to assess the likely effects of a contemplated sequence of move commands and to uncover potential problems before actually sending the commands to the vehicle in space. Depending on their implementation and levels of integrated models, digital twins may be used in all stages of a thing's lifecycle, including design and build phases in addition to operation.

The concept of digital twins has been applied to IoT systems [4]. In its basic form, an IoT digital twin can be a data structure in the cloud that represents physical things, such as sensors and devices. Twins receive data and status updates from their thing counterparts and thus mirror the thing states as close to real time as possible.

IoT digital twins often contain representations of the actual as well as of the desired states of the device. Applications and services may modify the desired state to cause the twin implementation to issue the necessary commands to the device to modify its state accordingly and to bring the two into compliance. Cloud-based applications and services can use and access digital twins as device proxies instead of accessing the endpoints directly. The use of digital twins in IoT systems has several potential advantages:

- Faster access
- Always available
- Saves bandwidth and power
- Abstract representation and interfaces

A cloud-hosted digital twin can be accessed by the cloud applications on demand and without the latencies involved in accessing the actual device or waiting for it to report. Obviously, this access obtains the last reported thing state which is not necessarily the current one. This might be OK in many cases since instant synchrony in distributed systems is not possible anyway due to inevitable communication delays.

The twin state representation is always available, even during periods of time when the device is disconnected or sleeping. This allows delay-tolerant applications to proceed without having to wait for the temporarily inaccessible devices.

Savings of bandwidth at the edge and on the way to the cloud occur when access to digital twins can substitute for direct access to devices. Bandwidth is usually in ample supply in the cloud where the digital twins reside, but it may be costly or restricted at the edge. Power may be saved in constrained environments by not forcing the device to wake up and power up its transceiver to send a report.

The abstract representation refers to the uniform data and command formats that digital twins can maintain for cloud applications to query thing data or to issue instructions to change a desired state. This approach can simplify the implementation and increase portability of applications by hiding device differences, intricacies, and details from the cloud applications. It is accomplished by using the twin implementations to translate between the generic representations and the device-specific data formats and commands.

In principle, IoT digital twins may be enhanced by incorporating device physical models and by combining their outputs with the real data to perform simulations, analysis, and predictions of thing behaviors.

Real-Time Stream Processing

Real-time stream processing operates on data in flight on its way from the origin to the consuming applications and services in the cloud. It captures the current state of the monitored IoT system. Various forms of streaming analytics may be deployed to detect system events and states that may indicate potential concerns or anomalies that need attention and may require immediate processing and reaction. This subsystem may also implement short-term storage to facilitate causal analysis of data that preceded events of interest.

In a normal system operation, much of the data tends to be within bounds and may not constitute significant events. Such data can be processed in relatively simple ways, such as being forwarded to the visualization, archived if necessary, or discarded.

Events in the incoming streams are forwarded to their intended destination, such as stream processors, the dashboard for visualization, and/or to direct operator alerting via messaging or email. More complex forms of real-time event processing involve detection of complex events and execution of functions and transformations that may involve multiple input streams and produce multiple outputs. This may be implemented as a messaging or publish-subscribe pipeline where various functions or microservices take input from one or more sources, transform it, and produce one or more outputs. Those outputs may be forwarded to other functional transformations, thus allowing the creation of complex flows. Various output stages may also be delivered to other cloud services and applications via data retrieval APIs.

The pipeline approach simplifies stream and event processing by providing modular processing functions and data connectors for external interfaces as well as the structured interfaces between internal blocks for composition of the more complex flows and functions. Stream processing frameworks and tools that are used for this purpose in IoT systems tend to be drawn from several application domains, including (1) embedded and control systems, (2) web/cloud, and (3) enterprise systems.

Simpler scripting tools and practices that originated in manufacturing control systems for PLC programming can be used for event processing in IoT systems. They are similar to those used at the edge, but in the cloud, they can operate with diverse data from multiple input domains. Examples of such systems include PLC-like scripting tools and flow-based programming environments such as the Node-RED discussed in chapter "Communications" in conjunction with the edge.

Systems originated in the web/cloud domain tend to be developed for massive web and Internet applications and services, such as the click-stream processing of browsing navigations and social network postings. They mostly evolved to add real-time and streaming dimension and processing capability to the massive batch systems, such as the MapReduce which is used to process Internet indexing piles to support the document search functions. As such, their major focus is on scalability and robustness.

One such stream processing system is Storm [5]. It uses pipes for data input streams, called spouts, and provides the means for fitting them into pipelines using the functional modules called bolts to form complex directed acyclic graphs called topologies. In effect, a topology acts as a data transformation pipeline. It is conceptually similar to the operation of a MapReduce job, but it operates in real-time as opposed to batch. A less elaborate option is provided in the Kafka publish-subscribe system that allows functional services to be interposed between its topics.

In the enterprise domain, there are tools for complex event processing (CEP) that support business rules management and decision-making, such as Drools [6]. These systems have rich functionality, but they tend to be resource-intensive and fairly complex to install and use. They are designed to meet the demanding enterprise volume and robustness requirements that may not be present in some IoT applications.

In general, event processing systems repurposed from other domains may be a good fit for IoT uses in terms of their functional characteristics. However, since they are intended for other uses, they may differ in terms of their assumptions on the nature of the incoming data and on the operational requirements such as the volume, throughput, scalability, and reliability. IoT system designers should be thoughtful in devising their event processing solutions and selecting the tools that strike the right balance between the convenience, complexity, overhead, and capability to meet the needs of their particular system.

Short-Term Streaming Storage

Some implementations of the IoT streaming processing path include a form of storage that provides fast access to the detailed recent history of data reports and events. The idea is to keep the high-precision data around for a while to be able to roll back and analyze significant events by examining the preceding relevant state transitions that may have caused them. Afterwards, data may be aggregated and down sampled for trend analysis and transfer to the longer-term archival storage.

Short-term storage for streaming data is intended to provide the means to query data by time and to replay recent events for uses by streaming analytics and similar applications. It is usually optimized for fast access, and its entries may be kept in memory or in the fast-access storage devices, such as the solid-state devices (SSDs).

This type of storage has comparatively short data retention times, on the order of hours or days. Its implementations tend to resemble large circular message buffers. Due to the absence of the archival requirement, they have much less structure and overhead than the traditional databases.

Cloud IoT Data Storage

IoT data and events ingested from the edge sources may be stored for batch processing and archival purposes such as the long-term comparisons or auditing. Computationally intensive applications, such as ML and AI, often work with large collections of data in the batch mode to create and refine inferencing models. Analytics, such as the building management, may use swaths of archival data such as the building telemetry on comparable days from the previous seasons and years to improve their optimizations and daily and hourly predictions.

Properties of IoT Data

IoT data share some properties of the big data, often referred to as four Vs – volume, velocity, variety, and veracity. *Volume* refers to the very large scale that may require storage to be spread and partitioned over a large number of nodes and managed as a distributed database with the related issues of consistency and convergence among copies operated in parallel. *Velocity* is the propensity to arrive at high speeds, either from sources sampled with high frequency or in the aggregate as a cumulative rate from large numbers of endpoints regardless of their individual rates. *Variety* means that data may come from many sources and in different formats, generated by machines and humans, including sensors, things, and various forms of Internet data and services. *Veracity* refers to the quality and reliability of data, including accuracy and provenance. IoT data may have errors and missing samples due to numerous reasons, including operating in harsh environments, in marginal conditions, with unreliable power and networks, exposure to security breaches, and the like. They often need to be cleaned up and checked for sanity and consistency somewhere in the data path between the capture and processing by the applications.

Furthermore, IoT data have some unique properties that impose additional requirements on their storage and retrieval. Some of the key ones include:

- Time-series data and metadata
- Semi-structured and unstructured
- Lifecycle management and data retention
- Data access patterns, writing, and retrieval

The term time-series is usually applied to the IoT data sampled and reported at regular intervals. This form of data acquisition may result in continuous streams of reports from specific endpoints and is sometimes referred to as the (data) channels. Irregular occurrences, such as changes of discrete states or detection of exceptional conditions, are reported as events when they occur. IoT measurements and events are commonly marked with time stamps at capture or as close as possible to their point of origin. Time stamps are used for data correlation and access and need to be stored in the database. IoT data also often include metadata in the form of loosely structured key-value pairs that need to be stored and be retrievable with or in conjunction with the related sensor readings and events.

In general, IoT data have characteristics of unstructured and semi-structured data in the sense that they have a great variety of formats and length of records, partly due to the fact that they may contain variable sets of metadata.

Depending on policies for acquisition and reporting, IoT data may contain a lot of entries that are of little archival value, such as numerous consecutive readings of ambient temperatures that are in range and with little or no change. Given the volume of data and possible storage restrictions, especially at the more constrained intermediate nodes, it is generally a good practice to support a definition and enforcement of IoT data retention policies. This can also facilitate the faster retrieval of retained data. For example, older data may be averaged and down sampled for archival purposes which allow affected raw samples of low value to be discarded.

In terms of storage access patterns, incoming IoT data tend to be time-series sequences; thus their writes tend to be sequential and mostly appends. Depending on the nature of queries, retrievals can result in random reads of individual values or in larger sequences for data defined by the time intervals or attributes, such as proximity in terms of location or domain membership. If the database implementation provides some degree of control of its mode of data layout, it is generally advantageous to the structure of the IoT data in storage in a manner that makes servicing of the common case of large queries efficient.

Types of IoT Databases

Cloud implementations of IoT storage services typically consist of an input stage that receives incoming data through posting and subscriptions to the topics to be stored and writes them in the database. On the output side, the storage service implements support for queries and APIs for data retrieval.

The central piece is the database itself. Cloud and web database systems can be very massive and span thousands of servers in multiple geographic regions. They often focus on capacity, scalability, and fault tolerance by means of partitioning and replication. Cloud databases can be designed for structured or unstructured data with stream or batch processing modes and queries for data retrieval and with various models of consistency and convergence among the partitions and replicas of data.

Selection of the type of database for IoT systems is made primarily based on its fitness for representation and querying of IoT-style data and metadata which tend to be variable-length collections of time-stamped semi-structured and unstructured data. Of course, an IoT database also needs to meet the system requirements in terms of capacity, throughput, scalability, fault tolerance, and reliability.

Relational Database Management Systems (RDBMs)

Legacy commercial databases are primarily designed to store and retrieve information with a high degree of organization, such as financial records. They are predominantly of the relational type called the RDBMS (Relational Database Management Systems). RDBMSs are characterized by a fixed row-column format reflecting the exact data schema of their entries. They tend to be optimized for transactional processing with implementations that support the so-called ACID properties – atomicity, consistency, isolation, and durability. This is necessary for financial applications where a transaction may need to debit one account and credit another with the same amount in a consistent manner, i.e., both operations are successfully completed, or neither is in the case of failures. This mode of operation needs to be supported in the presence of any combination of software and hardware failures, including loss of power. Those requirements lead to fairly elaborate designs and implementations. For example, transactional systems usually keep durable logs of all actions that constitute a transaction and carefully manage their reliable completion, using algorithms like two-phase commits. In case of errors, logs are unrolled, and partial actions are undone to maintain consistency. These techniques [7] serve their intended purpose well, but they introduce complexity and overhead that limits their throughput.

Due to differences in requirements, legacy RDBMS are generally not a good fit for the storage of IoT data. They tend to impose a rigid schema on data which is not a good fit for the variable semi-structured IoT data types. Moreover, any changes of format or addition of new types of sensors and things may require redesign of the schema to accommodate their data models and reconfiguration of the database. That is a slow and expensive process during which the system may need to be taken offline. In terms of performance, RDBMs tend to have comparatively lower throughput and scalability due to the transactional overhead. Another potential issue is that the commercial relational databases intended for enterprise applications often incur high licensing and maintenance costs. For these reasons, legacy RDBMs should be avoided in IoT applications. It may be tempting to use them for small pilot projects since they may be familiar and available; however this implies that a redesign and replacement will likely be necessary later when moving such systems into production.

NoSQL Databases

As indicated earlier, IoT data have almost the opposite requirements of RDBMs design assumptions – high volumes and rates (throughput), little and variable structure, and no transactional requirements. They are often serialized as variable-length arrangements of key-value pairs of readings and metadata with time stamps.

Databases suitable for IoT data storage generally belong to the category called NoSQL databases. The name comes from drawing the distinction from the traditional commercial databases designed for structured data and commonly queried using the variants of the Structured Query Language (SQL). NoSQL databases do not support transactional processing and guarantees which eliminate the associated overhead. They are typically designed for big-data and web applications with high data volume, scalability, and throughput requirements. They often support multi-node and clustered implementations with partitioning for parallelism, geographic distribution, replication for resilience, performance, or combinations of those.

NoSQL is somewhat of an exclusionary definition. It says what those systems are not, but it does not indicate what they are. Based on their structure and primary mode of operation, NoSQL databases tend to fall into the following major categories:

- Key-value stores
- Document-oriented databases
- Column-oriented databases
- Graph databases

Due to the nature of IoT data, the key-value and document-oriented databases are probably the most popular types for IoT uses. *Key-value stores* or databases logically consist of stored values, addressed by keys.

Document-oriented databases treat entries as documents that may be indexed by various keys. Documents in entries can be almost any text, with XML and JSON being quite common. In IoT applications, this allows for the direct storage of input data that are often formatted as JSON strings. Documents can be of variable length, which can be an additional benefit in IoT systems as new types of sensors with different message formats, and lengths are introduced.

Column-oriented databases tend to be used mostly for massive applications such as the web page index storage and processing, mostly batch, for search applications. They are designed for petabytes of data spread across thousands of commodity servers in multiple data centers, such as the Google BigTable [8] and its Hadoop open-source simile Hbase [9].

Graph databases use graph structures for semantic queries of data. They consist of collections of nodes and edges, with the edges representing the relationships between nodes. Graph databases can be useful for representing complex structures and relationships in IoT systems, such as the layout and connections among the thousands of components in an HVAC system of a large building. Such representations can be used to visualize, model, analyze, and optimize complex system-level behaviors, such as the flows of energy and heating and cooling fluids in a building.

There are quite a few open-source and some commercial offerings of NoSQL databases, but the field is somewhat fragmented and not yet settled. Some of the more popular ones that claim IoT fitness and use cases include MongoDB [10] and Couchbase [11]. Some of them are offered as managed services in the commercial public clouds. Naturally, almost every cloud provider also has their own proprietary version of a NoSQL that they promote for IoT uses as well. Some of them are discussed in chapter "IoT Platforms" in conjunction with IoT platforms.

Several new databases are being developed specifically to support the time-series data, including open-source InfluxDB [12] and OpenTSDB [13]. They are positioned as tools for collecting metrics for DevOps and for IoT and real-time analytics. Both have built-in support for storage and queries of time-stamped data, and support the addition of a variable number of key-value encoded metadata with entries.

In addition to selecting a database that is a good fit for their system's requirements, IoT system designers need to consider the stability and potential longevity of the product and the vendor. The size of the relevant installed base can be a useful indicator. Some systems have been developed for different uses, and they may be a good fit in terms of the stated objectives, but – as discussed with the eventing systems earlier – they may be built on different assumptions and dependencies. For example, the OpenTSDB system is designed to work in conjunction with the HBase, and both need to be installed for it to operate. Such issues can introduce additional overhead and management complexity in the implementation of an IoT system that should be weighed against the potential benefits.

Cloud IoT Analytics

IoT systems collect data by sensing a physical system in order to provide insights into its state and behavior and, when the appropriate services are available, predict and optimize its behavior through recommendations or direct actuation. Preceding sections covered elements of data collection, storage, and retrieval. The remainder of this chapter is devoted to IoT system analytics and machine learning and artificial intelligence techniques that are often used to implement the analysis and acting parts. System applications and services that obtain insights and turn them into actions may provide some or all of the following:

- Insights into system state and behavior
- Descriptive analysis
- Prediction of future behavior
- Prescription of a course of action to achieve the desired behavior

In order to generate insights and translate them into actions, an IoT system may include computer-based tools for descriptive, predictive, and prescriptive analytics that can be applied to reliably and consistently optimize system operation. *Descriptive analytics* answers the "what happened" question by categorizing the data and interpreting the resulting system state. In its simpler forms, it may identify the nature and potential causes of the events of interest, such as detecting a water pump failure and determining that it is the cause of the loss of cooling in an area of a building. *Predictive analytics* addresses that "what will happen next" question by estimating the system trajectory based on the current state, constructed system model, and past behaviors in comparable situations. *Prescriptive analytics* is the most sophisticated form of analytics processing, and it provides a "how to get there" course of action that will guide the system on a desired trajectory. It usually works in conjunction with the predictive analytics or provides some of the predictive elements in its design.

Traditionally, predictive and prescriptive analytics tend to be domain-specific – such as for building management or grid management – and often require additional customization and tuning for specific installations. This tends to make them costly to develop and difficult to make portable. The work on data interoperability, metadata structural annotation, and standardized data retrieval APIs holds some promise to alleviate the problem.

Visualization and Dashboards

While much of the current attention is on the computerized tools and services for the data analysis, it is important to keep in mind that many of these functions have traditionally been and continue to be provided by humans in the role of system operators. The ability to acquire sensor data in real time and to visualize the quantified system state on operator dashboards makes an IoT system valuable in the control systems managed by humans. Instrumentation and visualization with remote control is usually the first step in exploitation and an immediate tangible benefit of an IoT installation before the descriptive and prescriptive analytics become available and operational.

A *system dashboard* is a user interface that visualizes system states and allows the operator to perform some control actions. It typically has a top-level option to visualize overall the system state with selected operational and performance metrics. A typical dashboard shows the overall state of the system and allows the operator to focus on areas of interest for more detailed information, up to including readings from specific sensors. The system dashboard is also used to highlight components that are malfunctioning or operate outside of the desired range. Those may be marked with different colors and coupled with operator warnings and alarms that, depending on the severity, may require explicit operator acknowledgment. This allows the operator to explore areas of potential concern and take remedial action if necessary.

Trained operators can use such data-based insights to manage the system. Coupled with their knowledge of the system design and day-to-day experience with its operation, they can go a long way towards managing the system within desired functional bounds. Depending on their experience and familiarity with the system, operators cognitively perform some levels of descriptive analysis in terms of understanding system behavior and determining causes of some of the issues that may require attention and corrective action. They can also make predictions about the likely trajectory of system behaviors based on the current indicators and trends and preform the corrective control actions that they deem necessary to achieve and maintain the desirable states.

Operator-based control is a default current practice and reasonably effective in managing complex systems, such as the power grid and large commercial buildings. However, it requires a well-trained dedicated staff of operators that is usually not economically justified for smaller systems. In comparison, machine-based analytics may be cheaper to acquire and operate, can work reliably and consistently around the clock, improves over time, and can often perform as well or even better than the skilled operators. Techniques and methodology for implementing machine-based analytics and optimization systems are covered later in this chapter. The next section illustrates the use and potential value of machine-based analytics in comparison with the traditional methods.

Use of IoT Analytics: An Example

IoT systems with Internet reach and access to large sets of data, and interoperable data formats provide an opportunity and incentives to create portable and widely usable machine implementations of analytic services. In this chapter we outline the data science tools that may be used for that purpose in IoT systems. Before doing so, we provide an example of analytics at work side by side with trained operators in a real system.

Figure 4.3 illustrates measurements in a large commercial building as reported to and visualized by the building management system (BMS). It shows temperature measurements for the segments of the three floors (35, 38, and 40) of a large commercial building during the office hours in the summer (cooling) season. These were selected by the operator through BMS dashboard controls for visualization, in order

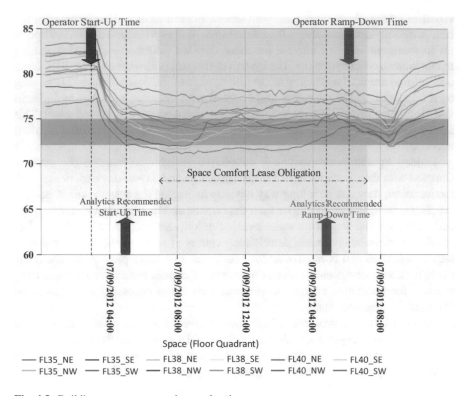

Fig. 4.3 Building temperatures and control actions

to monitor the areas of interest. Temperature measurements are reported in the degrees of Fahrenheit, ^{0}F. The green band of 72–75 °F (22.2–23.9 °C) indicates the desired target range of indoor temperature, and the extended band around it is the allowable range.

In order to control costs, the intensity of cooling is reduced, and temperatures may be allowed to drift out of the range comfortable to humans when the building is not occupied, i.e., before the arrival and after departure of the occupants. The time, when the more intense cooling starts to in order to normalize the building ambient temperature in time for occupant arrival, is often called the bring-up or the start-up time. It is usually commenced on a pre-programmed BMS schedule. It may be fixed for the season or determined by the system operators a day in advance, based on weather forecast, expected occupancy for a particular day, and prior experience with the building behavior. In this particular example, operators initiated the building start-up in the pre-dawn hours in the morning. Consequently, the ambient temperatures were reduced, and the building was in the desired operating range in time for the occupant arrival, nominally set for 7:30 am.

Towards the end of office hours, the reverse process of building ramp-down or bring-down was initiated to reduce the cooling in anticipation of occupant departure. This process usually begins some time earlier, because the building ambient conditioning has a degree of inertia that causes some lag in time between the change of settings and their measurable effect.

The blue zone in Fig. 4.3, whose endpoints are marked by the horizontal dashed arrow, labeled Space Comfort Lease Obligation, indicates the period of time when the building owner is obligated to maintain the agreed-upon space comfort conditions by the lease agreement. Failure to do so is subject to financial penalties.

During the scheduled office hours, many BMSs can manage the normal operation automatically by monitoring temperature thresholds and using the control scripts to adjust the cooling as and where necessary. This is akin to the autopilot mode for the BMS operation. When something unusual happens or a fault is detected, the BMS alerts the operators using the display indicators, notifications, or alerts commensurate with the severity of the condition.

In order to increase efficiency and conserve energy, system operators may provide a form of constant commissioning by providing the additional fine adjustments that commensurate with the building's changing conditions. For example, they may be able to reduce the intensity of cooling when the weather turns cloudy or the building occupancy drops below the levels assumed by the daily control scripts (say on a day before a long holiday when people may leave early). In addition, building occupancy can be dynamically tracked by sensors – such as the security gate counters – and used as guidance for the more responsive and efficient building energy management.

Control decisions that have a key impact on a building's energy consumption are when to initiate the start-up and ramp-down sequences. Some systems do this on a fixed seasonal schedule, which is inefficient. In energy-conscious operations (that reduce the energy bills and help the planet), operators make this prediction every day, a day in advance based on the weather forecast, occupancy assumptions, and past experience. These predictions are used to initialize the BMS scripts for the day and to execute and time the controls and activations of the HVAC system components accordingly.

Commencing the building start-up too early wastes energy as it reaches the space comfort conditions before the office hours. On the other hand, starting late can result in the failure to reach the comfortable condition in time, thus triggering financial penalties and making the occupants unhappy. The impact of this decision is amplified in many large cities where the energy is priced dynamically based on the time of day. The cost tends to peak and can be many times higher than the average in the morning hours when most office buildings are started up.

All of this is of special interest in our example, because the system depicted in the Fig. 4.3 was also running an analytics system in a mature state of development [14, 15] for comparison and testing purposes. This provided a rare opportunity to compare the operation of analytics with the actions of the trained professional operators familiar with the building. The analytics was designed to predict building behavior and suggest control actions to reduce energy consumption while maintaining occupant comfort. Its operation was based on a set of thermodynamic models, machine learning, weather data, and historical data on building's operation recorded for the past several years. The system was designed to provide operational predictions for 24 h in advance to allow adjustment of set points for the next day and also to provide short-term 2 h predictions for the more responsive constant commissioning during the operating hours.

For the purposes of our current discussion, building bring-up and ramp-down times recommended by the analytics for this particular day are shown by the vertical lines at around 5 am and 5 pm, respectively. The point to note is that later start-up and earlier bring-down times relative to the operator actions would have resulted in additional energy savings without impacting the occupant comfort.

Figure 4.3 provides an example of IoT optimization at work – a complex BMS system and analytics at work predicting start and stop times and guiding constant commissioning for efficient building management. Arguably not as spectacular as an AI program beating the world chess champion. But it is useful and saves energy and the related CO_2 emissions.

A detailed analysis of this building's operation with and without analytics demonstrated energy savings through a combination of optimizations of $505,000 during a measured winter season. A more thorough analysis of the portfolio of ten large commercial buildings located in New York City where the analytics algorithm was deployed for a year indicated average energy savings of 12% per building and on the order of $10 per square meter and approximately $5 M total [16]. This was accomplished in comparison with the state-of-the-art BMS and manual management by a team of skilled and experienced operators very familiar with those buildings.

This example provides a comparative study of a well-managed system with and without use of analytics and illustrates the practical potential of analytics to improve operational efficiency of control and IoT systems. The power of analytics is to continuously provide insights and guidance comparable or exceeding that of the trained operators. With the incorporation of machine learning techniques and accumulation of the historical data on the system under observation, it generally improves over time. An important aspect of historical data is to record predictions and outcomes, positive and negative, of the resulting controls and use that knowledge to additionally train and improve the analytics. With access to the large data sets from other comparable systems that IoT technology enables, it is possible to incorporate learning and effective strategies from a wide range of comparable systems.

Machine Learning and Artificial Intelligence

Machine learning (ML) and artificial intelligence (AI) have made great advances with notable successes in some applications, such as image classification and language translation. There is an expectation that these techniques will be useful for creating IoT applications that can provide deep insights and predictive and prescriptive analytics with little or no human involvement [17]. An early visible illustration of the possibilities is provided by the implementations of the autonomous vehicle operation. They use AI and ML algorithms to continuously process inputs from a multitude of sensors that characterize the vehicle state, its position, motion, and the surroundings to predict its desired trajectory and to issue the corresponding step-by-step control actions that safely move the vehicle in the real world.

Machine learning techniques use a novel and somewhat unprecedented approach of what amounts to having the computers learn how to perform certain tasks without being explicitly programmed to do so. This is a radical departure from the traditional computer science approach of providing detailed sequences of instructions for completing a particular task that a program is designed to do. Instead, machine learning works by having a computer evolve a working model for solving a particular task by presenting it with a set of training examples [18–21].

Machine learning is a technique that is at the core of the current AI systems and applications. Although there are not commonly adopted definitions of those terms, ML is generally regarded as being a part of and providing the toolkits for the broader field of data science. *Data science* is an interdisciplinary field that provides a methodology for extracting knowledge, insights, and predictions from data. It combines and advances techniques and theories from mathematics, statistics, information, and computer science [22].

Data mining, the process of discovering patterns in large data sets, is closely related and arguably a component of data science. Both data mining and machine learning provide modalities of fitting patterns to data and discovering relationships that may not be known to system operators and even to its designers. Traditionally, data mining tends to focus on discovering patterns in archived data, and thus it is more of a descriptive analysis tool, whereas machine learning tends to emphasize prediction. Individually and in combination, these methodologies provide valuable tools for harvesting the value of IoT data and facilitate implementation of descriptive and predictive analytics.

Artificial intelligence (AI) is a broad term that generally refers to computers performing tasks that would be regarded as intelligent if performed by humans. The field has been around for decades. After several largely failed attempts to mimic human cognitive processes, including early conversational and expert systems, AI has changed direction, adopted some of the machine learning approaches described later in this chapter, and achieved some spectacular results in solving some specific tasks.

It is a somewhat fascinating and radically different change that led to the resurgence of AI; it abandoned unsuccessful attempts to codify human intelligence. Instead, it turned to ML to enable computers to learn how to do something that humans have difficulty understanding and expressing in an algorithmic form suitable for the conventional programming approach. Its early successes were in the image recognition and classification. Those problems eluded the previous AI attempts to encode the cognitive aspects of visual processing but failed due to the lack of understanding of how the human brain performs those tasks. The use of the ML techniques, such as the neural networks structured in a manner inspired by the animal visual cortex (not its cognitive processing), proved to be effective.

Enabling computers to learn how to solve problems sounds like a great idea in principle. Why then not do all the programming that way? The answer is that, with the current techniques and tools, practical application of machine learning is often a difficult, painstaking, and time-consuming task. Application of machine learning requires specialized knowledge, a lot of data, considerable computational resources, and a lot of trial and error and works well only for a limited set of specific tasks. We discuss some of the intricacies involved in the process later in this chapter.

An inflection point in the public perception of AI technology occurred when it succeeded in besting human world champions in what are regarded to be complex games. IBM's Deep Blue defeated the then world champion in chess in 1997 [23]. Google's AlphaGo program and its successors beat the world's highest-ranked players in the game of Go on several occasions [24]. Another IBM AI system, Watson, won the TV trivia contest game of Jeopardy against the then highest-ranked human opponents by answering questions posed in a natural language [25]. More recent practical success occurred in the field of image recognition where AI models have surpassed humans in accuracy when classifying presented images as containing (or not) specific objects of interest, such as cars or door fronts. The typical error rate of human classifiers is around 5%, whereas AI algorithms have reached 3%. Moreover, AI algorithms have incomparable advantages in volumes and speeds of processing and can reasonably be expected to improve over time.

ML and AI generally operate and work best when they have access to the very large data sets, on the order of hundreds of thousands or millions of data points. As IoT data can be aggregated in clouds on a massive scale, there is a growing expectation that applications of ML and AI techniques may be able to produce similar results in IoT applications [17]. In theory, they could use ML and AI techniques to analyze, learn, and discover patterns in the complex integrated systems that may be too difficult or impossible to model using the traditional causal techniques. If successful, such systems could provide valuable guidance to operators by creating insights and by prescribing and automating operations resulting in the optimal behaviors of IoT systems. The grand vision is that such applications may be made portable and applied to managing a variety of IoT systems that might supplement much of the need for human involvement.

In the remainder of this chapter, we describe the principles, techniques, types, and uses of machine learning that may be applicable to IoT systems.

Uses of Machine Learning

In general, different types of machine learning structures and algorithms are suited for different applications, and one of the implementation challenges is to pick the right combination. Some of the more common uses of machine learning include continuous estimation and optimization, classification and clustering, ranking and recommendations, transcription and machine translation, synthesis, and data generation. They are briefly described in the remainder of this section.

Continuous estimation and optimization: In its simplest form, this can be a prediction of the next numeric value in a sequence, such as a value of some measured entity. More complex variants can produce actionable recommendations on managing control set points in a building or autonomously navigating vehicles on public roads.

Classification and clustering: The task is to categorize which of the k categories the input belongs to or to produce a probability distribution over a set of classes. Clustering may be used to identify groups that share similar characteristics, such as market segments. An image classification example consists of identifying if the presented image contains a specific type of a visual object, such as a car. A variant of classification of interest in IoT uses is anomaly detection. Its purpose is to categorize events that are out of the ordinary. For example, a combination of the rate and magnitude of a pump's current draw and vibration may be anomalous, which is of interest for failure prediction and maintenance alerts. A financial example of anomaly detection is to flag credit card purchases that are atypical and likely fraudulent.

Ranking and recommendations: Algorithms are mostly used in information systems, such as to rank order search results or products to be presented based on the relevance and possibly user's profile. Recommendation systems typically try to predict and suggest which product or movie a user may like (and buy) based on the patterns established by their prior selections and those of users with the similar tendencies.

Transcription and machine translation: Transcription usually means conversion of some unstructured input, such as handwriting or speech, into text consisting of a sequence of corresponding characters. Machine translation usually means taking as input a sequence of symbols in a written language and translating it into another. Combined with speech recognition, it can be used to translate spoken language into a text or into synthesized speech in another language.

Synthesis and data generation: ML algorithm's task is to generate new examples that are similar to those in the training set. As discussed later, this can be used to augment the training data set when sufficient real data are not available. Another use is for artificial rendering of landscapes and large objects in video games.

Types of Learning in ML

Machine learning operates by training a computer, more specifically the machine learning algorithm designed for that purpose, to finalize a processing model for a problem that it is not explicitly programmed to solve. This operates by using generally large data sets to train the model until it achieves a high degree of accuracy. In the context of this discussion, the word model refers to the machine learning model that is constructed in the learning process and is expressed in the internal format of the ML system in use. Once completed and validated, the model can be put to work by analyzing and making predictions on the data that it has not previously seen.

Based on their mode of learning and the nature of the training data, machine learning techniques may be classified in two broad categories, with a few variants:

- Supervised learning – the training data set includes observed inputs and outputs.
- Unsupervised learning – the training data set includes inputs and no outputs.

Supervised learning operates with training data sets that include both inputs and outputs of a system that is to be modeled. In an image classification application, the training set of photographs in a system with supervised learning would include appropriate labels for items to be recognized, such as a car or a street sign. The objective of training is to construct a model that reasonably accurately predicts system outputs or classification for new inputs, such as images, that it has not previously seen. The name supervised is based on a somewhat stretched analogy with a learning system where a teacher can indicate the outcome of the task, such as the appropriate label or a value.

In *unsupervised learning* systems, the training data set contains input data points with no corresponding outputs. For example, it may include a large set of images that are not labeled. Such systems may provide useful insights, such as categorizing input data based on features that relate them in some manner which may or may not be obvious to their users or designers.

Semi-supervised learning systems use training data sets that are partially labeled, e.g., sets of images where only some of them have been labeled.

Reinforced-learning systems have the added property that their outputs can be assessed for accuracy and fed back to the model for adjustments and refinements of the model. They usually incorporate feedback from the actual system in terms of observations of the actual output. This enables the model to reinforce settings that produce correct outputs and to de-emphasize or eliminate those that do not. In effect, the machine learning model learns by doing and improves its performance over time. For example, an image recognition system may be reinforced by checking correctness of the labels that it produces – such as that there was or was not a street sign in an image that the system labeled as having the sign. Such feedback can provide reinforcement of the model in the sense of amplifying the settings that produced good outputs and attenuating the settings that produce erroneous results.

In a simplified example of IoT use, the training data set may contain a combination of inputs, such as the chilled water temperature and the cooling fan speed, and an output of the corresponding ambient temperature in a room. The machine learning system could use a number of such data points in its training set and attempt to "learn" the system so as to be able to predict the ambient temperature in the future given the values of its inputs. In a classification example, the training data set may contain related measurements of a pump speed and the current that it draws, each labeled with the output as being anomalous or in the expected operating range. The classification system would then attempt to model the system so as to be able to classify a state as anomalous or expected when given the specific set of previously unseen inputs.

Artificial Neural Networks

Machine learning applications use a variety of techniques, many of which are suited only for particular types of problems. Artificial neural networks are among the most versatile in the sense that their variants are applicable to a broad range of machine learning problems across different uses.

Neural network implementations in AI and ML use a basic computational element that is based on a highly simplified model of a biological neuron. In biological systems, neurons provide both the connectivity and the processing of sensory information. Neurons and their connections are generally believed to implement perception and cognitive processes in the nervous systems and brains of animals. A neuron is a cell that receives, processes, and transmits stimuli through the electrical and chemical signals. A neuron typically receives inputs from other neurons through a number of branched extensions, called *dendrites*. A protrusion, called *axon*, outputs signals from a neuron to other cells by connecting to their dendrites. An average lobster has approximately 100 thousand (10^3) neurons and a human around 100 billion (10^9), with on the order of 10^{14} connections between them (synapses).

A simplified model of an artificial neuron commonly used in neural networks is illustrated in Fig. 4.4.

An artificial neuron has a number of inputs, two or more, depicted as circles labeled x_1 through x_n. In the first stage of the network, inputs are usually the features of the system to be modeled, i.e., its attributes that are considered to be significant. In multistage neural networks, these would be outputs of artificial neurons from the previous stages. Complex neural networks that model systems with multidimensional features can contain multiple layers with a number of neurons in each. Such networks consist of an input layer, one or more intermediate or hidden layers, and the output layer. Such multistage models are called deep neural networks.

Each input line has a parameter, a positive or a negative number, associated with it that controls the relative weight and contribution of the related input in the mapping function that determines the output. These parameters are usually called

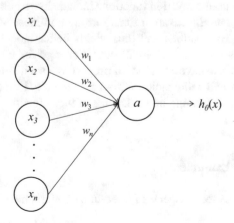

Fig. 4.4 Simplified model of an artificial neuron

weights in neural networks, and they are customarily labeled as w_1 through w_n, respectively. The input stage is usually called the input layer, which is followed by the activation layer and the output layer.

The central circle element, labeled a, represents an *activation function* that modulates the output of the neuron. Computationally, the output of an artificial neuron is a weighted sum of inputs to which a nonlinear activation function is commonly applied. It may be represented as

$$h_\theta(x) = a\left(\sum_{i=1}^{n} x_i w_i\right)$$

Features are some characteristics of the system and the phenomenon under study that are considered to be relevant for predictions or classifications provided by the machine learning algorithms. Features can be the raw data or derived by some processing or combinations of inputs, such as whether a bank customer had paid the rent or credit card balances in the past 3 months. Determining which features are relevant and how many of them need to be defined for the model to be effective is a very important aspect of implementation of machine learning. They may be defined through the process of feature engineering or discovered by data mining or machine learning.

The output, $h_\theta(x)$, is essentially a hypothesis function that commonly represents a prediction provided by the neural network – based on its topology and parametrized by weights – for a set of inputs presented to it at a specific point in time. In this equation α is an activation function. Neural networks typically use nonlinear activation functions, such as the logistical or sigmoid function that is defined as follows:

$$g(z) = \frac{1}{1+e^{-z}}$$

The shape of the sigmoid function is depicted in Fig. 4.5.

Other commonly used activation functions include hyperbolic tangent and rectified linear unit. In general, activation functions are nonlinear which often provides good functional approximations and thus yields better models. They tend to be smoothing functions that produce small output change for a small input change. Activation functions also have a number of properties – such as being differentiable, having a finite range, and being monotonic – that generally lead to better representations and shorter model training times. Details of those functions are outside of our current scope.

A Neural Network Example

This section illustrates some aspects of operation of neural networks by means of an example.

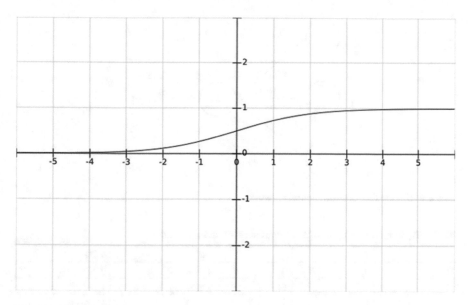

Fig. 4.5 Sigmoid function

Figure 4.6 depicts a simple feed-forward neural network with two inputs and the third one, x_0, used as a bias. The *bias* is an optional but quite frequently used neural network input that is always set to the value of 1. Thus, it acts as a constant whose value is defined by the associated weight. The weights assigned to individual inputs in this example are −5, 10, and 10, respectively. For simplicity, the two variable inputs in this example, x_1 and x_2, are assumed to take on only values of 0 or 1.

As per our earlier discussion, the output function for the neural network depicted in Fig. 4.6 may be expressed as

$$h(x) = \alpha \left(x_0 w_0 + x_1 w_1 + x_2 w_2 \right)$$

In this example, we use sigmoid as the activation function α. As indicated by Fig. 4.5, its outputs approximate 1 for positive inputs that equal or exceed 4 (0.982) and may be approximated as 0 for the inputs smaller than −4 (0.179). Thus, computation of $h(0,0)$ when both inputs are 0, becomes

$$h(0,0) = g(-5+0+0) = g(-5) \approx 0$$

and when the two inputs both equal one, the evaluation becomes

$$h(1,1) = g(-5+10+10) = g(15) \approx 1$$

All possible combinations of inputs are shown in Table 4.1. It shows that output is a matrix formed by applying the activation transformation to the weighted sums

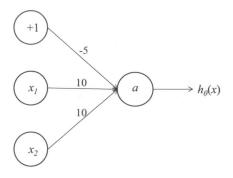

Fig. 4.6 Simple feed-forward neural network

Table 4.1 Inputs and outputs of the neural network in Figure 4.6

x_0	x_1	x_2	a	$h_\theta(x)$
1	0	0	$g(-5)$	0
1	0	1	$g(5)$	1
1	1	0	$g(5)$	1
1	1	1	$g(15)$	1

for the respective inputs, with the bias x_0 always set to 1. Incidentally, this neural network acts as a logical OR function.

A more elaborate neural network is depicted in Fig. 4.7. It contains an additional intermediate layer. In the neural network terminology, this is referred to as the *hidden layer*. It is hidden in the sense that it does not have direct external inputs nor does it produce direct external outputs. Neural networks with one or more hidden layers are called *deep neural networks*. As before, the input layer consists of just unmodified external inputs, called features. Those inputs are processed by the activation functions in the hidden layer, labeled α_1 and α_2 in our example, before being submitted as the weighted inputs to the next layer with its own activation function.

Inputs and structure connected to the first intermediate neuron, α_1, are identical to the neural network in Fig. 4.6 that produces an output equivalent to an OR function of its two variable inputs. By performing an evaluation similar to the one in the previous example, it can be shown that the last-stage neuron produces an AND function, and the complete network in Fig. 4.7 acts as a logical XOR function, $\oplus$, of its two inputs, which may be expressed as:

$$x_1 \oplus x_2 = \left(x_1 \vee x_2 \right) \wedge \neg \left(x_1 \wedge x_2 \right)$$

This example was constructed to illustrate the operation of feed-forward neural networks. The fact that the sample network computes a familiar function is atypical. It was constructed to indicate that neural networks can compute some familiar functions, albeit using a different approach and mode of operation. Except for the illustrative purposes, there is not much value in using the artificial neural networks to implement the known functions. The real power of neural networks lies in being able to predictably model patterns that do not have obvious or known

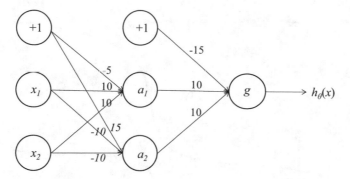

Fig. 4.7 Deep feed-forward neural network

causal counterparts or that humans do not yet know how to express in the algorithmic form. In most cases, the trained operational models of neural networks may not reveal any explainable causality between their inputs and outputs. Finished models may look like just a bunch of neurons, weights, and activation functions whose logic, if any, may seem inscrutable.

In fact, one of the major criticisms of machine learning is that it cannot be explained how some of the ML models reach their conclusions. They seem to magically solve the problem in a way that we may not understand, but it just seems to work. However, inability to explain how a model operates in causal terms can be a matter of concern leading to an understandable reluctance to rely solely on the unexplainable machine learning systems in some IoT applications.

An Example of the Neural Network Use

Having seen the basic operation of an artificial neural network, we illustrate how they may be used by means of an example. Let us assume that we want to solve a classification problem for the states of a particular system. This may mean designating each state, defined by the system characteristics and a given set of measured inputs, as belonging to one of the two classes, say normal or anomalous. The problem arises when we do not know or cannot algorithmically express the nature of the causality and the set of conditions that results in normal or abnormal states. Given sufficient data, this problem may be solvable by using machine learning.

Our task is to devise and train a neural network that will model the target system and be able to correctly classify its states. The network may be trained with the collected set of inputs. Depending on whether or not the resulting states are labeled, we may use supervised or unsupervised learning. The purpose of the network and the system model that it embodies is to be able to provide the proper classification in production. That means providing the classification of future system states based on the provided inputs, including (and especially) those that it has not seen before.

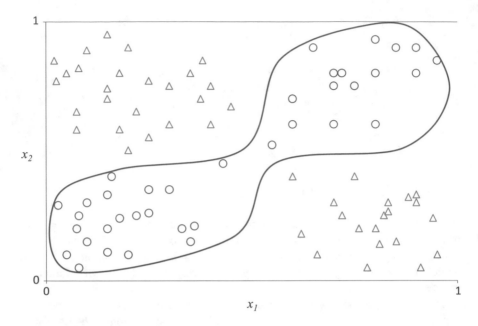

Fig. 4.8 Sample data set and separation curve

The details and steps of ML model development and training are described later in this chapter in the Putting it All Together section. For the purposes of this example, we fast forward the process to the point of having a sample data set and a trained neural network. In order to simplify the exposition, we drastically reduce the dimensionality of the problem to having only two inputs and two possible output classes.

Let us assume that we have collected a labeled data set depicted in Fig. 4.8. It is a scatterplot of two inputs, x_1 and x_2, that are normalized for simplicity and generality to range in values between 0 and 1. The outputs are the two categories or classes, one represented by the small triangles and the other one by the circles.

For computational purposes, we can assign values of 0 and 1 to distinguish between the two output categories. Since the primary intent is to provide classification, it does not really matter which label is assigned to which category, so let's assume that data points denoted by the small triangles belong to the category 0 and the small circles to the category 1.

The purpose of our classification neural network is to output the classification category for each presented data point, based on its attributes expressed as the values of its features and on the system model that it arrived at in the training stage. The system model embodied in the neural network conceptually has to behave in a manner as if implementing a function that separates the two categories.

We need a neural network that has two inputs and produces an output of 0 or 1. Since our problem appears to be nonlinear, it is reasonable to use a deep neural network. We can start with one hidden layer which is simpler and also happens to be a recommended practice. The network needs to have two inputs, our feature set, and provide a classification output of the category in which it is classified, 0 or 1, for each. We have created such a network in the previous example, so let us select that network configuration to be trained for this problem.

Weights are generally determined by training the model. Given that we know outputs for the training set, i.e., which category each data point belongs to, the neural network could be trained using supervised learning. The training would proceed by using the training data set, say a subset of entries depicted in Fig. 4.8. Let us conveniently assume that the training process produced the exact same weights as depicted in Fig. 4.7. This would create a production model that we could apply to classify outputs for the future inputs.

As an intuitive indication of why this might work, let us review the previously analyzed operation of our neural network and the labeled data set depicted in Fig. 4.8. The network produces a 0 classification (triangles) when either of the input features x_1 or x_2 is close to 1 and the other is close to 0. The network classifies the input data as category 1 (circles) when the two features have similar values, i.e., both are close to 0 or 1. In this simplified example, it is possible to confirm that this works by observing that data points belonging to the category 0 and marked as small triangles tend to cluster in the upper left and lower right corners of the scatterplot in Fig. 4.8. Numerically, this corresponds to the feature values close to (0,1) and (1,0). Conversely, data points belonging to the category 1, and represented by the small circles, tend to cluster around feature values close to (0,0) and (1,1). The activation function pushes the output to a decided extreme of either 0 or 1 which is what is needed in classification.

The ML algorithm is supposed to categorize inputs based on the specified features into the two categories. Conceptually, it should act as if implementing the nonlinear separation function outlined in Fig. 4.8. In practice, it should provide the correct separation, but the shape and nature of the curve may not be deducible from the structure of the neural network and the weights of its trained model.

This example was constructed to illustrate the basic operation of neural networks. In practice, neural networks are at their best when implemented on large and very large data sets with potentially numerous features. They tend to map into complex multidimensional spaces that can be difficult if not impossible to visualize or reason about in a manner that we did with our simple illustrative example. But the principle of their operation is the same.

Types of Neural Networks

Neural networks described in the previous section belong to the category of feedforward neural networks (FFNs). They propagate data through layers strictly in the forward direction until the output is produced. FFNs have one input and one output layer and may include one or several hidden layers of neurons, in which case they are also called deep neural networks and their learning process is described as the deep learning. There are many other types of neural networks that are designed to address some of the shortcomings of FFNs when used in the specific types of applications.

In this section we briefly overview several commonly used variants of artificial neural networks and their uses.

Recurrent Neural Networks (RNNs)

RNNs include loops whereby the output of a neuron at a given point in time may be fed back to be combined with the input at the next point in time. Output of a neuron is formed as a combination of its input from the data pipeline and some fraction of its previous output from the feedback loop. In this process, a neuron's weights and bias inputs generally remain unchanged. This modification in effect allows the creation of neural networks that have memory of the previous outcomes that can impact their processing. Other types of neural networks assume that input data are independent. RNNs are designed and well suited for processing sequences of events that are related, such as handwriting recognition, speech recognition, and language translation. A variant of RNNs, called the long short-term memory (LSTM) networks, introduces an additional input that can be used to remember (or forget) how much of a computed value to retain in the cell, for how long to remember it, and when to "forget" it. This can be used to modulate the length of the input sequence to be processed and commensurate with what the network is expecting, based on the nature of the problem and its processing capacity.

Convolutional Neural Networks (CNNs)

CNNs have connections between neurons inspired by the animal visual cortex. They are primarily used for tasks related to visual perception. CNNs are three-dimensional networks of neurons that structurally and operationally emulate the animal visual cortex in which neurons correspond only to the stimuli in a small segment of a visual field called the receptive field. CNNs have multiple layers that are functionally specialized into a convolutional layer, a pooling layer, and a fully connected layer. Input data, usually from images, are divided into smaller segments that are fed to the corresponding clusters of neurons and processed in stages. The rectified linear unit activation function is applied between the convolution and the pooling layers to highlight the features of potential interest. The fully connected layer then computes and outputs the class score indicating the likelihood of belonging to known visual categories, such as a car or a cat. This process is repeated for all segments of the image, and the cumulative result is aggregated for a final classification of likely objects in the image.

Generative Adversarial Networks

GANs are based on the somewhat different approach of having two neural networks compete against each other in order to improve the performance of both. They are often used for data augmentation, to increase the size of training data sets by producing artificial data. For example, image classification is often performed by CNNs that require very large data sets for training in order to become effective. If the available initial data set is not sufficiently large, a GAN may be set up to increase it by generating synthetic images. This basically works by setting up two neural networks. The first is the generator of synthetic images that it randomly intermixes with real ones and presents to the second network, the discriminator. The discriminator tries to distinguish between the synthetic and the real images and outputs a classification of real or synthetic (0,1). The two networks are then presented with the actual result, and each of them uses its mistakes to learn how to improve its operation – better fakes for the generator and higher detection rates for the discriminator. When the two networks reach a satisfactory balance, the generator network may be used to create synthetic artifacts of a particular nature that are useful for increasing the data set.

Classical Machine Learning Techniques

Besides the artificial neural networks, several other techniques based on more traditional methods are used in machine learning. Machine learning techniques initially evolved from the statistical processing and optimization work. This was accomplished by introducing the notion of training to derive a model that can predict outcomes for previously unseen data with reasonable accuracy. Statistical techniques were modified to focus on prediction rather than on the traditional confidence intervals. Focus on optimizations moved from generating distributions for known data sets to trying to improve prediction performance on the entire data set based on the analysis done on the (available) training set.

ML techniques based on modified traditional methods continue to be used due to their familiarity, comparative simplicity, and – in many cases – the ability to explain their results. Some of them are described in the remainder of this section.

Linear Regression

In machine learning linear regression is often used for continuous estimation or prediction of outcomes in time-series systems. It works by fitting a model to the training data and using it to predict outcomes for the new data. Model fitting is done as in the traditional linear regression, by devising a function that is the best fit to the data points in the training set. The model may be a first-degree polynomial, basically a line, or a curve defined by a higher-order polynomial. Functional expression of the model is often referred to as the hypothesis function because that is what the model uses for prediction of outcomes, based on the belief that it is the best possible representation of the system under observation. Fitness of the model is determined by calculating differences between model predictions and the actual observed outcomes in the training data set. This is called the *error* or the *cost function* of the model, and it is usually computed as the mean square error between the model predictions and the actual observed system outputs for the available data points. The objective of the training is to minimize the error. This is usually done by calculating the gradient of the cost (error) function. A gradient of zero indicates no error, i.e., the prediction equals the actual output. Training progresses by changing the model parameters in the direction that reduces the gradient until it reaches zero or some low value after which no further improvements are possible.

Some common issues with regression include underfitting and overfitting. *Underfitting* occurs when the model is too linear to capture the system behavior properly. *Overfitting* refers to situations where the model learns and fits training data too well, resulting in poor performance on the general data set. Manifestations of underfitting and overfitting may also arise in ML models based on artificial neural networks.

Logistic Regression

Logistic regression has its roots in statistical modeling where it is primarily used to predict the value of a dependent variable. This is often the probability of some binary outcomes, such as the equipment failure or the risk of developing a particular disease. In machine learning, it is often used for classification by predicting which category an event belongs to. It uses modeling techniques similar to linear regression but with a significant difference that the prediction line or a curve is regarded as a division boundary that separates different categories or classes. Classification output is class membership (small number of discrete values) based on which side of the division boundary a data point lies on.

In logistic regression, inputs are features of the system under observation, multiplied by coefficients called predictors that ML adjust by learning to fit the model of the system with the goal of minimizing prediction errors.

Support Vector Machines (SVMs)

Support vector machines (SVMs) belong to a class of learning algorithms that perform linear classification of data into two categories. Their distinctive feature is mapping of representation of data points into a space that separates classification categories, with as wide and clear gap between them as possible to emphasize the division. SVMs can also be used for nonlinear classification by deploying the so-called kernel trick that maps input into higher-dimensional feature spaces in which a hyperplane may provide good separation. They can also be used in unsupervised learning to act as clustering algorithms.

Bayes ML Algorithms

Bayes machine learning algorithms are based on the application of Bayes' theorem, also called Bayes' law, that states how to compute the conditional probability of an event given that another event, that may be related to it, has occurred. Naïve Bayes classifiers are a quite common variant of this approach in machine learning that makes a strong independence assumption (hence naïve) between the input features. Extensions of Bayesian classifiers into networks with optional tree augmentation have also been applied to solve regression problems in machine learning.

Clustering and Decision Trees

Clustering in machine learning generally refers to placing data into one of the several distinct categories or clusters with similar features. There are several variants of clustering algorithms that differ based on their mode of operation. A popular K-means clustering is based on defining centroids, technically central cluster vectors and intuitively centers of gravity of individual clusters. The number of clusters, k, needs to be specified in advance which is one of the drawbacks of this algorithm. It often works by randomly initializing centroids and iterating through the training data set, possibly in multiple runs with different initial values, to minimize prediction errors. Variations of the algorithm are sometimes given different names. For example, one of them that starts with an initial assignment where data points can belong to more than one cluster and is referred to as the fuzzy clustering. Clustering can also be performed by assuming that data in a cluster likely belong to the same statistical distribution, often Gaussian. This gives rise to distribution-based clustering. Connectivity-based or hierarchical clustering is based on the notion that related data points are closer to each other in some space and thus operate on clustering them based on some measure of distance expressed as a distance function.

Decision trees can also be used in machine learning as predictors of continuous variables (regression trees) and classifiers (classification trees). Random forests are a variant that incorporates collections of decision trees.

Putting It All Together – ML Model Creation

Designing and training of ML and AI models can be quite an elaborate and time-consuming process. Steps include training that results in the formation of the model and validation to assess its fitness and correctness, as measured by its

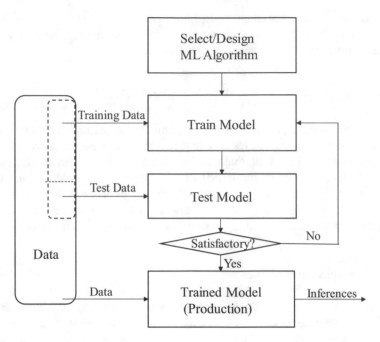

Fig. 4.9 ML model development

ability to predict outcomes. Completion of these steps finalizes the model so that it can be used in production to process the archived or streaming data, as appropriate, and to make predictions and classifications of the previously unseen data, which is its intended purpose. In IoT practice, this is usually preceded by a trial period to make sure that recommendations of the model are beneficial to the system and improve its effectiveness, such as in the example presented in Fig. 4.3. The trial stage may also be useful to gain confidence of operators and process managers.

The process of model development requires many decisions, including the type of and architecture of the algorithm to use, selection of the training and test data sets, performance evaluation metrics, methodology, and others. When there are issues with performance, and there often are in the initial stages, decisions involve whether to gather more data, do more training, and change the model architecture or even the algorithm.

Key stages in the model development process are depicted in Fig. 4.9 as a fairly orderly progression for illustrative purposes. As is often the case when designing complex systems, the process is iterative, and some of its parts may be revisited and revised, sometimes with the cascading effects and impact on the others. A general rule and the best known practice are to start with a reasonable algorithm approximation and to iterate through the process guided by the purposeful experimentation and attentive monitoring of performance indicators.

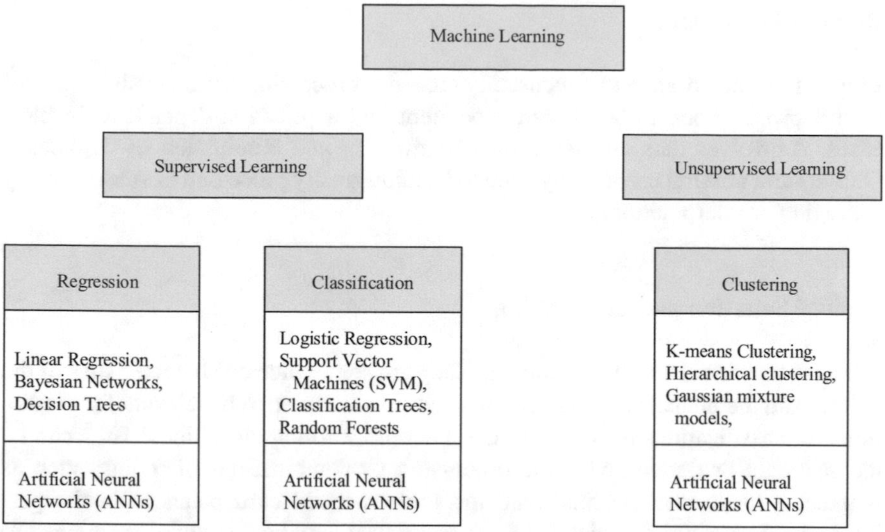

Fig. 4.10 Major types of machine learning algorithms

Key steps involved in the ML model development are depicted in Fig. 4.10, and some of their aspects are described in the remainder of this section. They commonly include:

- Data collection
- Preparation of data – cleaning and filtering
- Feature engineering – selection of relevant attributes
- Model selection and design
- Training
- Testing and validation
- Use (production) – prediction and classification

Data Collection

Collection of data, more is usually better, is the obvious first step. Data sets typically contain inputs or features and some outputs that may be labeled or unlabeled. Labeled data include some annotation of their attributes, such as a normal or abnormal state of a component or an image that contains particular types of objects. In IoT systems, data include telemetry, events, and inputs and outputs that may be acquired from other sources, such as people's inputs and comments, weather forecasts, and energy generation data. They may also contain historical data, including derived data and records of past behaviors, predictions, and outcomes. Data may be annotated with metadata during input or in post-processing by adding labels, automatically or manually.

Preparation of Data

Preparation and curation of data usually precedes processing and analysis. Specifics of this process tend to be system-dependent and application-dependent. In most cases, it involves data cleansing such as filtering and elimination of duplicates. Outliers and missing entries may need to be additionally processed or substituted by averaging or interpolation.

Feature Selection and Engineering

Inputs to the model are called features. They are some selected characteristics of the system and the phenomenon under consideration assumed to be relevant for predictions or classifications provided by the machine learning algorithms. They can be the raw data or derived by some processing or combinations of inputs, such as whether a bank customer had paid the rent or credit card balances in the past 3 months. Determining which features are relevant and how many of them need to be defined for the model to be effective is a very important aspect of implementation of machine learning. They may be defined through the process of feature engineering or discovered by data mining or machine learning.

Feature engineering refers to the selection of attributes and variables, called features of interest, that may exhibit correlative and predictive characteristics when processed by the model. In some cases, the choice of features may be obvious, and in others it may involve an elaborate manual or mechanical process called feature engineering.

Data points can have multiple features of potential interest. For example, in order to assess the probability that a borrower will pay back a loan, a prediction model may consider a number of features – such as age, gender, marital status, income, debt, net worth, credit rating, owns or rents the home, and the like. The task of feature engineering is to identify which features are most relevant for predicting the outcome and selecting those as inputs to the model. Intuitively, more features may lead to the better performing models. On the other hand, a large number of features can lead to more complex, cumbersome, and slower to execute models. Being able to reduce features to the most relevant subset is the task of feature engineering. This can be an iterative process that applies different metrics and analyses to determine the likely impact of specific features. The intent is to eliminate the ones that are redundant or not relevant, and – where possible – compute some new derived features that may represent the influence of several others and can be strongly correlated.

Feature engineering is somewhat of a black art. It can be a demanding and tedious process [26] that may involve manual processing by humans. There is ongoing research in automating this process [27]. The amount of effort spent on this activity depends on the complexity of the problem and on the production importance of the resulting model. It may be worthwhile to indulge in when creating the important long-running production models, such as the bank loan approval or language translation. In other cases, one might start with an informed guess and reiterate if needed to improve the model performance. In general, models based on neural networks tend to be adaptive and may perform reasonably well with less investment in feature engineering than some of the simpler models based on the traditional techniques.

Algorithm Selection and Design

One of the key decisions in implementing an ML model is the selection or design of the algorithm to use. As shown earlier, specific algorithms tend to be well suited for the specific types of tasks. Choice of the algorithm depends on several factors, including the nature of the task to be performed and, to a certain extent, the nature of the available data set. For example, supervised learning requires availability of labeled data sets.

Figure 4.10 summarizes the major types of machine learning algorithms and techniques. The boundary between the supervised and unsupervised learning in practice is not as strict as the figure suggests, and some techniques and methods can be used with either approach. As indicated, ANNs are the most versatile tool in the sense that they can be used to implement all major types of ML applications which largely accounts for their popularity. Classical methods are sometimes simpler to train and execute, but they tend to work well only for limited sets of applications. They should be considered as an implementation option in cases when they are a good fit for the requirements of a particular problem.

A common approach in ML algorithm selection is to pick an existing algorithm of the type suited for the nature of the task that it needs to perform. As Fig. 4.10 indicates, there is a variety of possible choices with different characteristics and implementation complexity. Once the algorithm category that defines the type of the model is selected, it is quite common to start with one of the many models available on the Internet, many of which are open sourced [28–32]. This allows for quick prototyping and potentially faster convergence to the final model. Moreover, many of those models are already trained for the specific uses and may thus contain the resulting weights and parameters. Even when not an exact fit, they can be a good starting point for experimentation with a repurposed model. For example, to construct a model for object recognition of a specific type, such as the location of basketball players on the court in a TV broadcast, one might start by picking a generic algorithm, likely a convolutional neural network, that works well for detecting people in images and then tune it for the specific category of interest.

A more demanding route is to design a custom algorithm. This should be avoided when possible, as it requires costly ML expertise and development time and can turn out to be a time-consuming research project that may or may not yield the expected result in the time available. One of the perhaps non-intuitive practical problems with ML in general and custom models in particular is debugging. Namely, when a model in development is misbehaving, it may be difficult to determine if the cause is incorrect implementation or a wrong model, since the "expected" behavior is not known.

Either way, once the algorithm is chosen, it may be necessary to select or specify a particular architecture. In ANNs this usually means choosing the number of layers and the number of neurons in each, called the depth. For example, the good rule of thumb, the first preference in FNN design, is that one hidden layer may be enough. Deeper networks can use fewer neurons and have fewer weights to train per layer, but they generally tend to be harder to optimize.

Model Training

Once the ML algorithm with the desired architecture is in hand, it needs to be trained until it is able to achieve the required performance. For this purpose, portions of the available data are set aside for model training and validation. Data to be used are divided into a *training set*, a *test set*, and optionally a *cross-validation set*.

The training can start with some random assignment of weights that gets modified in accordance with the models' successes and failures in categorizing data from the training set of inputs and outputs. It is quite common to start the training with some preliminary assignment of weights, such as those found in the reference model or even a random selection that needs to converge to the proper settings in training. The process then continues by varying the settings in accordance with successes or failures when processing the training data set.

An important aid in determining the convergence and effectiveness of the model is to define performance metrics and to track it in the course of model development. Performance metrics can include accuracy, error rates, and various other indicative trending indicators. Many ML libraries and toolkits provide mechanisms for doing so. For example, the Scikit-learn Python library at one point provided 14 variants of metrics for classification systems, 8 for clustering and 6 for regression [33]. The tools can often calculate and plot the selected performance indicators as the model is running through the training iterations. They can be observed and used as guidance for evolving the model in the directions of increasing accuracy and reducing errors. They may also be used when the model is in production and operation, especially during the initial validation period.

One of the techniques often used to increase the learning speed and tuning of neural networks is the *back propagation*. This refers to backwards adjustment of weights in relation to the estimation errors in the training set. The basic idea is to tune the weights of inputs to artificial neurons in proportion and direction that reduce the magnitude of the error [34]. Predictions are evaluated for accuracy at the output, and the corresponding weight corrections are applied to the inputs of each stage, propagating backwards from the output towards the preceding neural layers. Hence the term backward propagation, as it refers to the direction of flow of adjustments which is opposite to the normal processing flow.

In back propagation and model development in general, adjustments are based on calculating the multivariate derivative of the error (cost) function, its gradient.

The objective is to follow the direction of lowering gradient in the process of minimizing the error which terminates when the minimum is reached. In practice, error curves can have saddles with local minima that may not be global, and additional tests may have to be performed to detect the difference if present. Another issue that may be encountered in practice is the *vanishing gradient*, where gradients in the last stages tend to be small and their derivatives become increasingly smaller when applied to the preceding layers in deep neural networks. This tends to happen when activation functions that produce outputs in the (0,1) range, such as the sigmoid or the rectified linear unit, are used in the output stages. Another practical problem is known as the *exploding gradients*, where larger initial values tend to grow excessively and can even reach nonrepresentable values in the process of back propagation. More detailed treatment of these issues is beyond our current scope. However, practitioners of machine learning have to be aware of them and deal with the related problems should they arise.

The ML model is trained until it starts achieving promising or satisfactory results on the training data set, as per performance indicators and the chosen metric. If training set performance is poor, usually the model may need to be improved. This is often done by iterating and changing some settings – such as the model capacity, architecture, or training time – guided by observing performance indicators.

Testing and Deployment

When the model starts converging and showing reasonable success rates, it may be tested with a cross-validation data set to validate its effectiveness on a different data set from the same system and to verify that it is not overfitting the model to just the training set.

When the model is performing satisfactorily on the training and the optional cross-validation data in terms of satisfactory accuracy and low error rates, it should be tested with the test data. As mentioned earlier and indicated in Fig. 4.10, test data are drawn from a different segment of the data set to reduce the cross-correlation with the training set. The results are supposed to be indicative of the model's performance with previously unseen data in actual operation. When the model performs satisfactorily with the test data, it can be moved to the intended productive use. If performance is poor with the test data, the remedy may be to gather more data and to repeat the training process.

Upon successful completion of testing, the finalized model can be deployed for production use. Once in operation, it is advisable to continue monitoring the model's performance by deploying various forms of performance metrics, including recordings of predictions and the actual outcomes of the resulting actions. Combined with the accumulation of the historical data from the system in operation, these inputs can be used to continuously tune and improve the model and its effectiveness over time.

Performance Optimization

Getting an ML model to the operational state often requires significant computational resources and time for the model training and testing. While model development costs are usually front-loaded and may take on the order of weeks, it also makes sense to optimize the production version as it may be in use for a long time and benefit from speed improvements. Operating the model efficiently can result in lower cost and/or higher execution speed for delivery of inferences and predictions when running. Another reason for optimization may be to enable the model to run in resource restricted environments, such as the edge nodes and things. An actual decision to spend time on either or both optimizations depends on the specific circumstances.

Hardware and runtime optimization options include tuning the implementation to run on the target hardware platform, such as the specific CPU and OS combinations. Some hardware manufacturers do that for a select set of the popular ML libraries [35]. All other things being equal, picking a supported combination of hardware and a library might be beneficial for performance. One of the important factors in the speed of execution turns out to be the type of arithmetic used by the model implementation, such as the fixed (integer) or the floating point. Some specific versions of hardware perform much better one way than another. In general, the fixed-point variants tend to execute faster.

Another option is to use general-purpose graphics processing units (GP-GPUs). These are the specialized hardware units originally developed as hardware graphics accelerators for gaming applications. They contain a large number of processing units, hundreds or thousands of cores, that operate in parallel with high memory bandwidth. They tend to be suitable for some common ML operations such as the matrix multiplication and can greatly reduce training and inferencing times. Model training or execution on GPUs may require use of special software tools, such as the GPU-specific programming library. Fortunately, they are becoming widely available and integrated with commercial platforms and AI offerings, such as the TensorFlow. The GPU hardware is available for purchase on commercial graphics cards and in the specialized AI cloud configurations. They are also available on the pay-as-you-go basis on the commercial cloud platforms.

More recently, field-programmable gate arrays (FPGAs) are being considered as hardware accelerators. These are chips with a large number of logic gates, memory, and interconnections that can be programmed in the field for any specific application. Some versions also include integrated CPU for general-purpose tasks and to feed the gate array. Functionally, FPGAs enable the creation of the customized hardware configurations optimized for the execution of specific tasks, such as a particular ML model. This has a great potential for optimization in uses that justify the added cost of custom development, usually performed using hardware description languages (HDLs) that require specialized expertise.

Another performance enhancement possibility is the use of hardware specifically designed to accelerate model development and execution stages. As discussed in chapter "IoT Platforms" in conjunction with IoT platforms, designers of one such specialized ASIC, the TensorFlow processing unit, report 30–80 times performance improvements over the comparable CPUs and GPUs [35].

In terms of function placement, the ML algorithm development tends to be performed in the cloud due to its demand for large data set and computational resources. Algorithm execution, however, may be performed centrally in the cloud or in distributed fashion with some inferencing done at the edge. As discussed in chapter "The Edge", edge models can have reduced complexity and perform basic inferencing by processing local data. Edge instances of the ML algorithm typically perform event detection on the local data and forward them to the cloud for aggregation and deeper analysis.

One design aspect of ML algorithms that can facilitate their functional decomposition and distribution is the pipeline approach. Namely, ML algorithms are sometimes implemented in stages as pipelines, to break the problem into smaller pieces and to enable different models to work on the different aspects of the more complicated problems. For example, an application was developed for Google maps to read street numbers and names on store fronts in images collected by the street-scanning vehicles to improve accuracy and automate annotation of locations. It was created as a combination of two ML models – one to detect the rectangular shapes with appropriate aspect ratios that are likely to be signs that may contain text and another one to interpret the text and numbers in those when present.

Moreover, edge models can collect and report performance statistics, such as success and failure ratios, to the cloud for refinement. The improved model may then be used to update its other edge instances and thus improve the overall system performance. This approach can also increase system throughput by operating many distributed instances of models in parallel.

Explainability of Model Results

As discussed in this section, many operational issues may be encountered in the process of ML model development and training. Even when all of those are successfully overcome and a well-functioning model is developed, a major issue of model explainability, or unexplainability of its results, remains.

As described earlier on the example of neural networks, an ML model is determined by its topology, parameters or weights, and activation functions. The weights are determined through learning, and they generally do not reflect any discernable descriptive pattern or causality for human interpretation. A bigger philosophical implication may be that the AI algorithms seem to operate by discovering some recognizable but unarticulated patterns (tacit knowledge), and not by the deductive reasoning and mathematically or algorithmically expressible causalities (codified knowledge).

Be that as it may, one of the major criticisms of neural networks is that the nature of their functioning may not be subject to interpretation and scrutiny. This can present significant problems, for instance, when used by the financial institutions to determine whether an applicant should be granted a loan. If a person is rejected, there is no way to determine exactly why and no easy way to re-examine or appeal the decision.

The use of unexplainable AI algorithms should be approached with caution in IoT applications that can impact the real world, especially in closed-loop automation settings that operate without a human in the loop. At the very least, such uses should be preceded by extensive periods of testing in an advisory mode, where the system gives suggestions for human operators to consider and evaluate before initiating actions. This process also creates operator familiarity with the system operation, which makes it more likely that they may be able to intervene to remediate problems when mishaps occur. Otherwise, operators may be faced with a very difficult and uncertain task when asked to intervene and fix an automation malfunction without knowing how the system got to that particular state. The situation may be even worse when the problem is the result of an adversarial action or a security breach.

References

1. Armbrust, M. et al, (2009) 'Above the clouds: a Berkeley view of cloud computing', *Technical Report UCB/EECS-2009-28*, University of California Berkeley, [Online] Available at: https://www2.eecs.berkeley.edu/Pubs/TechRpts/2009/EECS-2009-28.pdf (Accessed Dec 15, 2019)
2. Jonas, E. (2019) 'Cloud programming simplified: a Berkeley view on serverless computing', *Technical Report UCB/EECS-2019-3*, University of California Berkeley, [Online] Available at: https://www2.eecs.berkeley.edu/Pubs/TechRpts/2019/EECS-2019-3.pdf (Accessed Dec 15, 2019)
3. Kafka [Online] Available at: https://kafka.apache.org/ (Accessed Dec 15, 2019)
4. Tao, F. et al (2019) 'Digital twin in industry: state-of-the-art', *IEEE Transactions on Industrial Informatics*, 15(4), p 2405–2415.
5. Storm [Online] Available at: https://storm.apache.org/ (Accessed Dec 15, 2019)
6. Drools Business Rules Management System [Online] Available at: https://www.drools.org/ (Accessed Dec 15, 2019)
7. Gray, J. and A. Reuter (1993) *Transaction processing: concepts and techniques* San Mateo, CA: Morgan Kaufmann
8. Chang, F. et al. (2008) 'Bigtable: a distributed storage system for structured data', *ACM Transactions on Computer Systems*, 26(2), p 4:1–26.
9. HBase [Online] Available at: https://hbase.apache.org/ (Accessed Dec 15, 2019)
10. MongoDB [Online] Available at: https://www.mongodb.com/ (Accessed Dec 15, 2019)
11. Couchbase [Online] Available at: https://www.couchbase.com/ (Accessed Dec 15, 2019)
12. InfluxDB [Online] Available at: https://www.influxdata.com/ (Accessed Dec 15, 201
13. Open TSDB [Online] Available at: http://opentsdb.net (Accessed Dec 15, 2019)
14. Wu, L. et al, (2012) 'Improving efficiency and reliability of building systems using machine learning and automated online evaluation', *2012 IEEE Long Island Systems, Applications and Technology Conference (LISAT)*, Farmingdale, NY.
15. Anderson, R. et al (2014) 'Di-BOSS: research, development and deployment of the world's first digital building operating system' in Capehart, B. L. and Brambley, M. R. *Automated diagnostics and analytics for buildings*. Fairmont Press, Inc., Lilburn, GA, p 109–125.

16. M. Rudin et al. (2015), 'Buildings finally get a brain: Di-BOSS' in Wood, A. and Gabel, J. *The future of tall: a selection of written works on current skyscraper innovations*, CTBUH, Chicago, IL. p 82–91.

17. Chui, M. et al, (2018) *Notes from the AI frontier* McKinsey Global Institute [Online] Available at: https://www.mckinsey.com/featured-insights/artificial-intelligence/notes-from-the-ai-frontier-applications-and-value-of-deep-learning (Accessed Dec 15, 2019)

18. Goodfellow, I., Bengio, Y. and Courville, A. (2016) *Deep learning* MIT Press, Cambridge, MA. [Online] HTML version available at: http://www.deeplearningbook.org/ (Accessed Dec 15, 2019)

19. Nielsen A. (2015) *Neural networks and deep learning* Determination Press [Online] Available at: http://neuralnetworksanddeeplearning.com/index.html (Accessed Dec 15, 2019)

20. Aggarwal, C. C. (2018) *Neural networks and deep learning* Springer International Publishing, Cham, Switzerland

21. Ng, A. (2019) 'Machine learning' Online Course, [Online] Available at: https://www.coursera.org/learn/machine-learning (Accessed Dec 15, 2019)

22. Berman, F. et al (2018) 'Realizing the potential of data science', *Communications of the ACM*, 61(4) p 67–72

23. Weber, B. (1997) 'Swift and slashing, computer topples Kasparov' *The New York Times*, May 12, p1, 20.

24. Sang-Hun, C (2016) 'Google's computer program beats Lee Se-dol in Go tournament' *The New York Times*, March 15.

25. Markoff, J. (2011) 'Computer wins on Jeopardy: trivial it's not', *The New York Times*, Feb 16

26. Panigrahy, P.S., Santra, D. and P. Chattopadhyay (2017) 'Feature engineering in fault diagnosis of induction motor', *2017 3rd International Conference on Condition Assessment Techniques in Electrical Systems (CATCON)*, Rupnagar, p. 306–310.

27. Kanter J. M. and K. Veeramachaneni (2015) 'Deep feature synthesis: towards automating data science endeavors', *2015 IEEE International Conference on Data Science and Advanced Analytics (DSAA)*, Paris, p. 1–10.

28. Scikit [Online] Available at: https://scikit-learn.org/stable (Accessed Dec 15, 2019)

29. Pytorch [Online] Available at: https://pytorch.org (Accessed Dec 15, 2019)

30. Keras [Online] Available at: https://keras.io (Accessed Dec 15, 2019)

31. Mahout [Online] Available at: https://mahout.apache.org (Accessed Dec 15, 2019)

32. Tensorflow [Online] Available at: https://www.tensorflow.org (Accessed Dec 15, 2019)

33. Scikit model evaluation tools [Online] Available at: https://scikit-learn.org/stable/modules/model_evaluation.html (Accessed Dec 15, 2019)

34. Rumelhart, D. E., Hinton, G. E. and Williams, R. J. (1986) 'Learning representations by back-propagation errors', *Nature*, 323, p 533–536

35. Elmoustafa, O. et al (2017) 'Tensorflow optimizations on modern Intel architecture' [Online] Available at: https://software.intel.com/en-us/articles/tensorflow-optimizations-on-modern-intel-architecture (Accessed Dec 15, 2019

Chapter 5
Security and Management

IoT installations can be complex distributed systems with many geographically dispersed heterogeneous nodes with different capabilities, connected via separate physical networks. They can span multiple operational and security domains managed by different entities. Securing such systems poses challenging problems.

While security is a concern in all computer systems, the problem is even more acute in IoT systems because they can interact directly with the physical world and can thus impact the health and safety of people, as well as the physical and operational integrity of manufacturing equipment and processes. Security breaches of IoT systems can disrupt power generation and distribution, disrupt manufacturing processes in a way that cause harm to people and the environment, endanger lives of medical patients, and take over control of autonomic vehicles and numerous other scenarios with serious effects. In 2012 Leon Panetta, US Defense Secretary at the time, warned that the USA was vulnerable to a "cyber Pearl Harbor" from an aggressor nation or an extremist group that "… could derail passenger trains loaded with lethal chemicals. They could contaminate the water supply in major cities, or shut down the power grid across large parts of the country." [1]

Another important and elevated design consideration in many IoT systems is the preservation of privacy. Privacy in IoT systems is an issue that goes well beyond common concerns, because monitoring of the physical world can reveal sensitive information including the whereabouts and the activity of users in real time by direct observation or by inference. This can pose property and safety risks, such as making burglars aware when premises are unoccupied and even unlocking doors to intruders.

While consequences of IoT security breaches can be dire compared to those encountered in IT-only systems that do not interact with the physical world directly, some aspects of IoT systems set them apart and help to reduce the attack surface, focus security measures, and mitigate risks. For one, IoT Edge and many other components are fixed or narrow-function devices, not general-purpose computing systems. IoT systems, whose endpoints tend to act as servers of data and commands, do not need to be constructed in the Internet fashion of allowing a large population of

© Springer Nature Switzerland AG 2020
M. Milenkovic, *Internet of Things: Concepts and System Design*,
https://doi.org/10.1007/978-3-030-41346-0_5

users with various degrees of endpoint security protection to access and interact with them. Instead, IoT devices can be directed to communicate only with a limited population of known authenticated entities that may be identified and with whom bidirectional trust can be established. Moreover, additional precautionary measures may be taken to reduce exposure, such as not allowing the downloading of unsolicited or unauthenticated software, closing of ports not used for IoT communication, and elimination of OS features that allow remote login, shell access, or support unsafe protocols such as Telnet.

This chapter discusses security requirements in IoT system and common techniques for establishing, monitoring, and maintaining it. It starts with the discussion of threats and vulnerabilities and two examples of documented breaches. Subsequent sections cover risk and threat analysis that informs security design. The section on cryptography describes principles and algorithms that are the foundation for many of the contemporary authentication and security protocols. The chapter continues with the discussions of security mechanisms and mitigating techniques including endpoint security, network security, and monitoring. The section on management covers the node life cycle and steps involved in bringing it into an IoT system and managing its operational states. The last section provides an overview and summary of how all of these components can be made to work together in designing and implementing security in an IoT system.

IoT Security Threats and Vulnerabilities

At its most basic, security may be viewed as protection from unauthorized access or changes of an IoT system. In this context, change can mean anything from data falsification to malevolent actuation that can cause harm to people and the environment. Security is a system-wide property, and a combination of techniques at different levels may need to be deployed to maintain it and to detect and mitigate effects of breaches should they occur.

As implied by its name and definition, an IoT system many include connectivity and potential for access from parts or all of the public Internet. This makes it potentially vulnerable to most if not all of the known types of cyberattacks associated with the Internet and distributed systems in general. Moreover, IoT systems may provide additional exposure when operating constrained things at the edge that communicate over wireless networks deployed in settings that are not physically secure and might be accessible to intruders. Constrained wireless nodes and networks tend to support less robust forms of security, thus making them more susceptible to eavesdropping, spoofing attacks, and physical tampering.

The primary objectives of security in information systems are to protect confidentiality, integrity, and availability of information. They are sometimes referred to as the CIA triad, an acronym formed from the first letters of their names.

- Confidentiality – ensure that information (data) is available only to authorized parties.
- Integrity – ensure that information (data) is accurate and unchanged.
- Availability – ensure that information is available for access when needed.

Points of potential security exposure in a system are sometimes collectively referred to as *the attack surface* that adversaries may try to exploit. Virtually any part of an IoT system may be subject to an attack, including the sensor layer, short-haul (wireless and wired) communications, gateways and fog nodes, long-haul communications (wired and wireless), cloud, applications, and services.

Types of security attacks generally fall into several broad categories, including:

- Physical attacks.
- Software attacks.
- Network attacks.
- Encryption attacks.

In the remainder of this section, we briefly overview some types of attacks that are relevant to IoT systems [2].

Physical attacks are possible in almost any setting, but IoT systems with nodes "in the wild" that use wireless networks can be particularly susceptible. Physical access to a node can make it vulnerable to tampering, and even replacement with a malevolent version of the node that may be functionally similar and fool the rest of the system into believing it is a genuine one. For this to happen, a fraudulent node would have to obtain access to the security protections, if any, used by the replaced authentic node. This may be possible to accomplish by hardware tampering to obtain the node's identity or cryptographic material or by software tampering through malware code injection. Even lower-tech modes of physical tampering can cause system malfunction, such as using a gas lighter to cause a thermal sensor to generate a faulty temperature reading or placing a still picture of an innocuous scene in front of a surveillance camera to obstruct its view and conceal a nefarious activity. Even less sophisticated, but potentially effective, might be to sever the camera power cord to render it useless.

Software attacks, such as malevolent code injection, can enable various forms of undesirable altered behaviors, such as commanding an actuator to perform unsafe actions in the physical world. Another common form of attack is denial of service, where an adversary can flood the system with traffic that causes an overload of either the processing nodes, communication links, or both and thus makes the system unavailable to legitimate users to carry out their normal operation. Another form of weak security, especially in the consumer space, is the configuration of nodes with the default common user, password combinations. Such systems may be relatively easily breached by obtaining those or by simply trying the common defaults or the ones used by the node manufacturer if known. One such attack is described later in this chapter.

Network attacks can compromise an IoT system in a number of ways. They include various forms of eavesdropping and spoofing, such as "the man in the middle attack" where an adversary virtually inserts an attack node between two IoT nodes and intercepts, possibly alters, and replays messages that they are sending to each other without realizing that they are communicating with an imposter. Such attacks can involve disclosure of information, capture of credentials or cryptographic information, and disruptions of the system by inserting malevolent messages or commands.

Wireless networks can provide additional exposure because they can be eavesdropped without requiring physical taps on the communication medium. Listening to a discovery broadcast can reveal the identity of the sender and possibly other nodes that respond to it. Some constrained devices rely on network security and implicitly regard as legitimate any communication on it. Some networks have low protection, such as using of a shared encryption key which can give practically unlimited access to an intruder who obtains it. Even without breaching their security, wireless networks can be disrupted by causing interference or by jamming the radio frequencies on which they operate.

Routing is another potential attack surface in mesh wireless networks. One such attack called the sinkhole attack is for a node to declare itself as having exceptional resources and power, thus causing its neighbors to choose it as a routing waypoint and direct their multi-hop traffic to it. Another variant, called the wormhole attack, involves two malevolent nodes conspiring to claim that there is only a single hop between them and thus divert a lot of routed traffic to themselves. Once they obtain legitimate messages, adversarial nodes can disrupt the system operation by inappropriately terminating or altering messages that they are supposed to be routing.

At the application or services layer, the well-known attacks include downloading of code infected by the malicious worms and viruses and phishing to steal user's credentials. In general, they should be less of an issue for the core of an IoT system, since its nodes should be designed to not allow such interactions with external initiators, especially the ones that are not authenticated.

Operational Technology (OT) Security Considerations

As described earlier, IoT systems may incorporate parts or the entire installations of legacy industrial and manufacturing systems, often referred to as the Operational Technology (OT) systems. The security design in such systems has to cover the combined needs and all of its constituent parts, including the OT and the Information Technology (IT) parts.

OT systems have traditionally operated separately and in isolation from IT systems. They tend to be bespoke or highly customized systems running on equipment – including sensors, actuators, and servers – that commonly resides in physically secured facilities and uses dedicated communication networks. The security of OT systems is often based on the physical security and isolation of control systems and, implicitly or explicitly, on the obscurity of its design. The latter is attributed to OT system customization and custom coding often with proprietary equipment, runtime systems, naming, addressing, and communication protocols. The result is often deemed to be hard enough for the trained personnel to modify and all but impenetrable to outsiders. These assumptions have been proven wrong by the security breaches such as the Stuxnet worm described later in section.

Keeping the system up and running safely to perform the intended productive function is the primary operational concern of OT systems [3]. OT system operational and design concerns revolve around safety, reliability, availability, resilience, and integrity. Given potential physical harm resulting from faulty operation, OT systems prioritize safety to avoid or minimize danger to people and the environment by striving to avoid them and to contain the impact when mishaps occur.

Safety is handled by rigorous, design, testing, and certification of components and processes, coupled with continuous monitoring to keep the system operating within the target boundaries and tolerances.

Reliability may be viewed as the availability of the system to operate continuously during production times, excluding scheduled maintenance and planned down times. It may be improved through redundancy by installing and operating multiple instances of critical components in parallel and having the standby take over if the primary fails. Monitoring and use of analytics for failure prediction can lead to improved availability by guiding preventive maintenance and component replacement to avoid failures.

Resilience refers primarily to the ability to absorb and limit effects of manageable faults and to reconstitute and resume operations after major mishaps and accidents. It may also entail graceful degradation where the system continues some meaningful operation or is guided through safe state transitions even when failures diminish its capacity.

Proliferation of IoT and its many benefits are causing Internet-enabled functional overlays to be added even to legacy OT control systems for added services such as monitoring, asset control, and data analysis. This brings in at least partial connectivity to IT-flavored components and possibly the Internet and with it many of the security exposures that were not contemplated in the design, analysis, certification, or attestation of OT systems. Security design in hybrid OT/IT systems needs to meet the requirements and address concerns of both of those environments. It may also introduce the requirements of privacy and confidentiality that were traditionally not a major concern in OT systems.

Examples of IoT Security Attacks

Due to their exposure to vulnerabilities and potential business impact on the victims, details and even occurrence of possibly many cyberattacks are not well publicized. In this section we discuss two security breaches that are fairly well document and illustrate some modes of penetration and outcomes of successful attacks.

Stuxnet

The Stuxnet worm attack is an example of a sophisticated cyberattack aimed at an industrial control process. It was very advanced from an IT point of view and unprecedented from the OT point of view. It proved wrong the prior OT belief that control systems isolated from the Internet and accessed only by authorized personnel are secure from cyberattacks, and even more so if they run some niche or custom industrial software [4].

Incidents attributed to Stuxnet were reported as attacks allegedly against the Iranian uranium-enrichment centrifuges in the summer of 2010. Be it as it may, such malware once released tends to find its way to other intended and unintended targets. For example, Chevron was the first US corporation to admit that Stuxnet had spread across its machines shortly after its discovery in July 2010 [5].

The attacked network is believed to have been isolated from the Internet. The attack was implemented as a computer worm that was used to infect target computers via USB sticks that contained the malware. Computer worms are self-propagating executables. Unlike computer viruses, they do not need other host programs to attach themselves to.

The Stuxnet worm has the design that was apparently intended to damage industrial processes controlled by the SCADA software. SCADA installations are widely used to control manufacturing and many other systems such as smart buildings, oil and gas pipelines, water systems, power generation and distribution, wind farms, and airports. Reported incidents were for a tailored version that targeted the Siemens WinCC SCADA system running on Windows and used to program and monitor SIMATIC series of Programmable Logic Controllers (PLCs) [6] managing centrifuges from the two particular vendors.

Basic Stuxnet operational steps are:

- Infection – of computers via USB sticks, could be any computer on the target network or a laptop of an authorized employee.
- Propagation – propagate itself to other machines through any available network or data cable connections.
- Activation – check whether the machine is the target, i.e., running a specific version of PLC SCADA control software.
- Compromise – if target, compromise logic controllers using the elevation of privilege exploit.
- Control – watch target PLC to learn the control-program steps; when learned, subvert the control program to vary centrifuge speeds in a manner that causes them to fail.
- Deceive – conceal harmful behaviors by sending false status information to the central control and management dashboard claiming that operation is normal.

Stuxnet's other notable features included use of four different zero-day exploits, forged digital certificates (subsequently revoked by the issuer) that required use of private keys from legitimate vendors, use of the first known PLC rootkit, and concealment of its diversion by manipulating PLC status reports. All of these are unusual and interesting in their own way, but their detailed discussion is beyond the scope of this chapter.

The creation of Stuxnet required a lot of resources and expertise in crafting worm-based cyberattacks, such as knowing the internals and programming of the particular SCADA and PLC systems. Its reported use is unusual in the level of sophistication and in the narrow target focus, both of which are believed to be the work of nation states. It shows that a determined attacker willing to go to great lengths can compromise and damage control systems that are believed to be safe by virtue of isolation, controlled access, and obscurity which is still a rather commonly held view in the industrial OT systems. The practical reminder here is that total security is impossible and mitigation should include incident recovery. Therefore, in addition to measures to protect security, OT and IoT systems need to provide constant monitoring to detect anomalies and incident-handling procedures to confine the damage and to restore the normal secure operation as soon as possible afterwards. All of those are discussed later in this chapter.

Mirai Botnet

The large and growing installed base of consumer IoT devices that are generally not very well protected provides opportunities for cyberattacks at a very large scale. While constrained edge devices and networks generally have higher security exposures due to their limited capabilities, consumer IoT devices are attractive targets due to their sizable installed base, notoriously lax security, and rampant use of default login and password credentials.

A common method used by hackers to exploit these weaknesses is to compromise large number of devices and turn them into armies of bots that can be commanded to perform coordinated distributed denial-of-service (DDoS) attacks. The process consists of two phases. First, an IoT device is compromised and infected by malware that provides a back door for access to the remote servers operated by hackers. In the second phase, compromised nodes are directed to attack a specified target by initiating some action or requests for service that consumes the target's resources. Individual actions, such as requesting a TCP connection or issuing a query, are usually legitimate and benign by themselves. The harm comes when large numbers of infected nodes are coordinated in time to issue repeated requests. In aggregate, those requests tend to overload the target by exceeding its service capacity. This can render the target unavailable to provide service to legitimate requesters, thus effectively bringing it down.

A well-documented attack of this type that occurred in 2016 is known as the Mirai botnet [7]. Compromised devices included a variety of commercial IoT and Internet-connected devices, such as routers, IP cameras, and DVRs. The attack started by scanning a large number of pseudorandom IP addresses and subnets for devices with open Telnet IP ports, TCP 23 and 2323 in this case. When found, the attacking server would attempt a brute-force login with known default credentials and commonly used user ID, password login combinations. These were drawn from a list of 62 entries including combinations of admin, user, guest, root, support, tech, password, 12,345, 1111, and the like. If successful, the IP address and login sequence were reported to the attacking server. The server program would then log in; try to determine the platform type of the target, such as processor and OS; and install the corresponding version of the malware with the hardwired address of the attacker's command and control (CC) server. The CC server could then issue commands to its army of bots, infected nodes, to attack specified targets at the specified time.

The Mirai malware was able to infect 65,000 nodes in the first 20 hours following its launch, and its population grew to a steady number of 200–300 K nodes, with a peak of 600 K infections in November 2016. Infected nodes were disproportionately concentrated in Brazil, Columbia, and Vietnam accounting for 41% of the total. The targets were mostly in the USA and Europe, one more example of the global reach, power, and perils of Internet connectivity. The documented targets included a popular blog site on security, Dyn domain, online game providers, and several telco companies. Primary points of attack were IP addresses, followed by some specific subnets and domain names. The peak per-target bandwidth reached 600 Gbps, a rate never previously seen in similar attacks. Dyn provides managed DNS service to commercial customers, at that time including Amazon, Netflix, Twitter, GitHub, and several online game providers. It appears that games were the primary target, but their DNS lookups overwhelmed Dyn's service, which in turn knocked out the services of its other commercial customers.

Security Planning and Analysis

Security is an end-to-end property that should be designed in IoT systems from the beginning as attempts to add it on later usually result in inferior protection. In general, total security is probably impossible and certainly too expensive to achieve, so the pragmatic objective is to put up enough of well-placed barriers and roadblocks to dissuade the would-be attackers and make them look for easier targets.

The security design process usually starts with the risk analysis and threat analysis to identify key threats and potential consequences of their occurrence. This is followed by the selection of security techniques and procedures that offer the most effective protection against identified threats.

Risk Analysis

Risk analysis can be applied to a number of factors that may impact the effectiveness and the cost of implementing and operating complex installations and systems. It can be a rather formal process undertaken to identify the nature and likelihood of events that could adversely affect a system and to estimate their impact should they occur. In manufacturing systems, risks can include safety, impact on the environment, liability, brand image, and financial losses from damage to equipment and disruption of production. Cyber security can be a major factor in this process for IoT systems, as its breaches by malevolent adversaries can impact the equipment and the processes that it may control.

In general, risk cannot be eliminated so the objective is to identify key threats and concerns that should be mitigated. One of the outcomes of the risk analysis is to guide the choice of proper and balanced level of security system-wide that meets the operational objectives at a cost that provides an acceptable return on investment. For security, this means mitigation of identified risks with a systematic approach to software security implementation, monitoring, auditing, and patch management.

Risk avoidance seeks to eliminate the risk by eliminating exposure to the specific threats. One manifestation of it may be to consider elimination of nonessential system features that cause specific exposures. In practice, this decision should be balanced with the potential adverse impact of the de-featured system's usefulness.

In addition to mitigation and avoidance, risks may also be accepted. *Risk acceptance* basically means making an informed choice not to invest in preventive measures against a specific exposure if it is assumed to have a very low probability, comparatively high cost to guard against, and/or its impact is manageable. This, of course, can be a problem when assumptions used in the assessment prove to be wrong.

One form of risk management is to transfer it to a third party, such as an insurance company, in exchange for payment. This is not uncommon for low-probability events that can have a very high impact and mitigation costs, such as natural disasters.

From the security standpoint, *risk transference* can be the avenue of the last resort to protect against unanticipated incidents, but it is no substitute for a good design. Naturally, insurance companies will do their own risk assessment to estimate the amount of premiums, and those tend to be inversely proportional to the extent of the assessed risk.

Security Threat Modeling

After key concerns are identified by the risk analysis, a common next step in designing a security system is to create a threat model of attacks and threats that should be guarded against. Threat modeling is not a formal discipline, but it is a useful process to go through in the early stages of an IoT system design. The primary objective of a threat model is to identify how an attacker might attempt to compromise the system and to make sure that the related mitigations are put in place. It should guide decisions for selection and implementation of countermeasures for identified vulnerabilities that must be eliminated or guarded against.

Like many other aspects of system design, this process is iterative and intertwined with other design activities and decisions. It is usually preceded by some level of the risk analysis that identifies vulnerabilities that must be avoided, guarded against, and perhaps some that may be given lower priority. Threat modeling also needs to take into account the setting in which the system will operate, including whether it is a physically secured environment, and factors such as the availability or lack of the network and power infrastructure, especially for the constrained edge nodes.

The basic steps in the threat-modeling process are:

- Model the system, create an architecture diagram (components and flows).
- Enumerate threats.
- Mitigate threats.
- Validate mitigations.

Threat modeling is usually performed by analyzing the system model. It is a good system design practice to create a model of the system early on and to revise it as necessary to reflect its changes and evolution. The model should be available as a part of the overall IoT system design. It should identify all key components, their connections, and data and control flows between specific components and in the overall system, along the lines discussed in chapters "Introduction and Overview", "The Edge", and "Cloud". Its construction is primarily driven by the functional requirements that specify what the system needs to do, such as monitor and control the operation of a manufacturing or a logistical process. This leads to a definition of the type, number, and placement of sensors and actuators and processing functions that need to be performed on them to achieve the stated purpose. The model is evolved by determining the functional placement of individual hardware and software components, the nature of their interactions, and the connectivity required to support them.

Once constructed, the system model can be used to enumerate potential threats to its components and connections. A good point to start is the list of known threat types. The commonly used Microsoft STRIDE model [8], named as the acronym of its elements, provides a helpful mnemonic checklist of threats as listed in Table 5.1. In practice, attacks may combine several of these modes and vectors, such as tampering with a node to reveal some aspects of its crypto system and/or identity and using those to impersonate that node for penetrating and disrupting other parts of an IoT system.

Depending on the nature of its components, connections, and the location and setting in which they will operate, some of these threats may not be applicable to the specific IoT system under consideration. Part of the job of threat analysis is to identify the ones that are.

When enumerating and analyzing threats, the general advice is to think like an attacker and to examine ways in which they may attempt to penetrate the system. This is useful as far as it goes, but the adversaries can be clever and resourceful and devise methods and forms of attack that the design team did not think of. A more systematic and complementary approach is to first identify the valuable system assets that an attacker may want to obtain and that you want to protect. It is also important to articulate and to keep track of the system design assumptions, especially

Table 5.1 STRIDE threat model

Threat	Property violated	Description
Spoofing	Authentication	Impersonating someone or something else, e.g., a device
Tampering	Integrity	Altering hardware, data (in motion or at rest), or code
Repudiation	Nonrepudiation	Denial of performing an action (inability to trace)
Information disclosure	Confidentiality	Exposing information to unauthorized parties
Denial of service	Availability	Deny or degrade services, e.g., through excessive resource consumption
Elevation of privilege	Authorization	Gain capabilities without proper authorization, e.g., system access

the ones that would appear to eliminate certain threats, and to make sure that they are actually implemented in the system as specified.

Once the vulnerabilities and threats are enumerated, the next step is to mitigate them by designing in the appropriate defense mechanisms, as described later in this chapter. One of the objectives of this stage is to identify system features that introduce vulnerabilities and to eliminate them by possibly omitting contemplated features that are not essential for system operation.

The final step in this process is to validate that the chosen mitigations address all identified vulnerabilities to the extent that is reasonable within the scope of the risk analysis.

Cryptography

Cryptography and cryptographic systems provide a foundation for ensuring confidentiality, integrity, and authentication of message exchanges in computer systems in general and in IoT systems in particular. They are the basis for encryption and securing applications and data in nodes and in transit. In this section we cover the basic principles of cryptography and some of their aspects that are relevant for securing and authenticating nodes and communications in IoT systems. Readers familiar with cryptography or not interested in its foundational details may skimp through this material. As indicated later, since the algorithms are public, the common practice in IoT systems is to implement cryptography using some trustworthy libraries. When selecting a particular scheme to use, it is important to strike the right balance between the security requirements, the relative strength, and the computational and deployment complexity of available choices.

The dictionary definition of *cryptography* is the enciphering and deciphering of messages in secret code or cipher. It is an age-old technique for exchanging messages securely between remote parties and keeping their content confidential even if intercepted by adversaries. Simple text ciphers, such as an early one attributed to Julius Caesar, worked by letter substitution. For example, a message may be encrypted by substituting each letter in a message with another one that follows it some fixed distance later in an alphabet. The history of cryptography and of the related code-breaking attempts is quite long and fascinating. However, in this chapter we focus on the more pragmatic aspects of cryptography that are relevant to computer systems and IoT in particular.

A cryptographic system usually consists of an encrypting and decrypting method, often combined with a key. Its basic components are illustrated in Fig. 5.1.

The system works by performing an encryption transformation on the plaintext of the message to be protected, shown as the plaintext P. An encryption key may be used for the encryption, shown as the encryption key ke in Fig. 5.1. This creates an encrypted message, C, called the *ciphertext* that is meant to be unintelligible to all unauthorized parties that may intercept or eavesdrop on it while in transit. The intended recipient needs to know the decryption algorithm and, if used, be in the possession of the decryption key to decrypt the message and to reproduce the original message as the plaintext.

Systems that use the same key for both encryption and description are called *symmetric cryptographic systems*. For this to work, an additional mechanism for secure key distribution is required. This may be accomplished by using separate channels, such as mail, or key exchange mechanisms described later in this section. In *asymmetric cryptographic systems*, such as the public-key systems, different keys are used for encryption and decryption.

Code breaking is referred to as *cryptanalysis*. Goals of an attacker may include deciphering of a particular message or preferably of the cryptographic system and the secret keys for decrypting all future messages among the parties that use them. There are three basic types of code-breaking attacks. The *ciphertext-only* attack occurs when an adversary comes in the possession of the encoded message. Other more dangerous forms include various combinations of the matched portions of plaintext and ciphertext, such as the *known plaintext* and the *chosen ciphertext* attacks.

Cryptographic systems are subject to brute-force attacks where the ciphertext can be computer-analyzed by rapidly trying many, and in some cases all, different combinations or keys. The process can be aided by exploiting known weaknesses

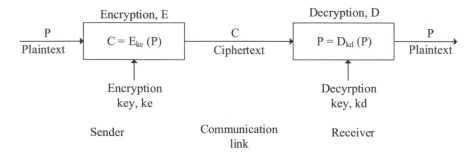

Fig. 5.1 Cryptographic system

and patterns, such as the frequency of words and letters in a given language. In modern systems, this disqualifies simple schemes such as the substitution ciphers and calls for the use of more sophisticated techniques and algorithms, some of which are described in this chapter.

The principle of open design is widely adopted in the security community meaning that contemporary cryptosystems do not rely on the secrecy of the algorithms but rather on the computational complexity provided by the robust algorithms and the secrecy of keys. This makes them open to the continued public scrutiny and probing by researchers who may document weaknesses and suggest ways for improvement and evolution. Moreover, public algorithms usually lead to commercial and open-source software and hardware implementations that can increase user confidence in them and can be used as building blocks to simplify an IoT system implementation.

Major security objectives of cryptography include:

- Confidentiality – protection of transmitted data for use only by authorized parties.
- Integrity – assurance that the data has not been tampered with in transmission or storage.
- Authentication – assurance that the data is from the claimed source.
- Nonrepudiation – assurance that the sender cannot deny sending the message.

Public-Key Cryptography

Public-key cryptographic systems are asymmetric and use different keys for encryption and decryption. Two types of keys are used in the process – publicly visible keys and private, sequestered keys. Each communicating entity uses a known method to compute a matching pair of its public and private keys. Public keys may be visible to all, and private keys are kept secret by each node and possibly stored in hardware-secured registers as described later in the section on endpoint security.

The process of secure message exchange in a public-key cryptographic system is illustrated in Fig. 5.2. The sender, A in this example, looks up the receiver's public (visible) key, denoted as vB, and uses it to encrypt the message. The resulting ciphertext can then be sent to the recipient over insecure communication links. The receiving node, B, decrypts the encrypted message using its private (secret) key, denoted as sB, and reconstructs the original message. Confidentiality is ensured since only the recipient B knows its private key with which the message can be decrypted.

Algorithms for public-key cryptography need to satisfy the following properties:

- Encryption and decryption algorithms are easy to compute.
- It is computationally easy to generate a pair of public key and private key.
- It is computationally infeasible to derive a private key from the public key.

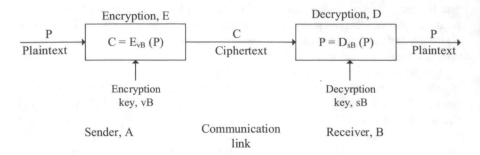

Fig. 5.2 Public-key cryptographic system

- Given a ciphertext and the public encryption key, it is computationally infeasible to reconstruct the original message.

Another optional but useful and rather common algorithm property is that either key may be used for encryption and the other one for decryption.

Using various combinations of message encryption and decryption, public-key cryptography may be used to meet the security objectives of confidentiality, integrity, authentication, and nonrepudiation. It can also be used for secure exchange of secret keys used by symmetric cryptographic algorithms.

When message is encrypted using the recipient's public key, *confidentiality* is ensured since the original message can be reconstructed only by the receiver that has the matching private key. An eavesdropper can obtain the ciphertext and access to the sender's public key but finds it computationally infeasible to reconstruct the original message.

Authentication can be accomplished by having the originator encrypt the message using its private key. Decryption of the message with the originator's public key uniquely authenticates the sender as it is the only possessor of the private key. Since it is impossible to alter the message without the sender's private key, this also assures its integrity. However, this scheme does not achieve confidentiality as anyone can decrypt the message using the sender's publicly known key.

In the previous exchange, the whole message serves as the sender's *digital signature*. A receiver can store the message and its plaintext equivalent and use it for nonrepudiation by presenting both the encrypted and the ciphertext versions to a third party, such as a judge, for verification. Since the message could not have been encrypted or altered by anyone other than the possessor of the encryption key, it can be uniquely associated with the specific sender who owns the related private key. Since processing and storing of lengthy messages, such as legal documents, can represent a significant computational and storage burden, digital signatures are usually implemented in practice by encrypting only a smaller part of the message, sometimes called the authenticator. It has to meet the requirement that it is impossible to alter the message without modifying the authenticator. A cryptographic checksum of the message may be used for the purpose.

Public-key cryptography may be used to simultaneously achieve confidentiality, integrity, and authentication by performing a double transformation illustrated in Fig. 5.3.

The sender, A, performs a decrypt transformation on the message using its private (sequestered) key, sA. It then encrypts the result using the recipient's public key, vB, and sends the message to B. B also performs a double transformation, starting by decrypting the received ciphertext to obtain $D_{sA}(P)$. This ciphertext is then encrypted using A's public key, vA, to obtain the original message. For this to work, the algorithm has to have the property that encryption and decryption can be applied in either order, that is.

$$P = E_{vX}\left[D_{sX}(P)\right]$$

The idea and principles of public-key cryptography were published in a paper by Diffie and Helman in 1976 [9]. The first major implementation of an algorithm satisfying public-key cryptography principles was provided about a year later by Rivest, Shamir, and Adleman [10], often referred to by the acronym RSA. The algorithm is based on the principles of modular arithmetic. It relies on the use of relatively prime numbers whose multiplicative inverses are very difficult to compute and, based on the evidence to date, satisfy the property of being computationally infeasible. More recently, elliptic curves are being used as somewhat simpler for computation while offering similar protection. Details of operation of these algorithms are beyond the scope of this book. To date, no successful breaking attacks have been reported for either scheme.

While simple in principle, and in comparison to the computation of inverses, public-key encryption transformations tend to grow in computational and storage complexity with the increasing message lengths and can be quite demanding. For increased security, use of longer keys is recommended, but it also increases the computational burden on nodes. For these reasons, in IoT systems, use of the public-key cryptography on less capable nodes is usually coupled with the addition of dedicated hardware assists as described later in this chapter.

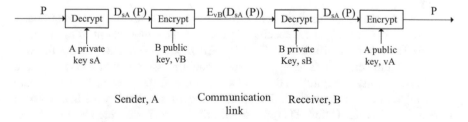

Fig. 5.3 Double transformation

Digital Certificates

While public keys are meant to be public, it is useful to provide some means of limiting their distribution and provide some degree of confidence that the public key, when obtained, belongs to a specific identified party. This may be accomplished by maintaining public-key directories administered by a trusted third party that protects the information and can exercise some form of authentication of queries. The directory approach may be useful within private IoT domains and groups of related nodes, but on a large scale – such as the Internet – it can become a performance bottleneck, single point of failure, and a target for hacking.

A more commonly used functional alternative for secure exchanging of keys is provided by the digital certificates. They are the basis for node authentication in most commercial IoT platforms as discussed in chapter "IoT Platforms".

A *digital certificate* provides a verifiable association between a public key and its owner. It involves a trusted third party, called certificate authority (CA), that issues certificates and vouches for their accuracy. In order to obtain a digital certificate, an interested party provides its identity and public key to the certificate authority. CA then validates the identity of the owner by some means, such as by examining the legal identity or corporate documents, and issues the certificate for the requested target entity, such as a domain or a specific server. A digital certificate contains the name of the entity that it is issued to, its public key, validity/expiration dates, and possibly some other optional information. The certificate is encrypted using the CA's private key.

Any recipient of a digital certificate may decrypt the information therein by using the CA's public key to determine the name and the public key of the owner. The use of CA's key and authority provides assurances that the certificate indeed originates from the named authority and that it has not been forged. This scheme works because only the authority can issue, revoke, or update certificates, say to extend their validity.

A node, A, wishing to engage in secure communication with node B, may request B to present its digital certificate to validate B's association with the provided public key. Optionally, B may also request A to present its certificate and provide the corresponding validation. Following mutual authentication, the two nodes may engage in message exchanges secured by the public-key cryptography with validated public keys.

A widely used format of digital certificates has been standardized internationally as X.509. It is a rather general specification with some simplified profiles that may be more suitable for constrained IoT nodes. An X.509 certificate includes the version, serial number, issuer's name (CA) and signature, issuer's public key, signature algorithm, validity or expiration date, subject name and ID (to whom the certificate is issued), and various extensions. The X.509 standard is defined by the International Telecommunications Union (ITU). Its use on the Internet is covered by the several RFCs [11].

Symmetric Cryptography

Symmetric cryptographic systems use the same key for encryption and decryption. They may be used to ensure confidentiality, integrity, and authentication of messages. Two communicating nodes wishing to use the symmetric cryptography must share a secret key. The sending node, A, uses a mutually agreed-upon encryption algorithm to encrypt the message using its copy of the shared key. This converts the plaintext message into the ciphertext. The receiving node, B, decrypts the message using the matching algorithm and its copy of the shared key. The cryptographic algorithm defines encryption and decryption as inverses of each other. Any other recipient or interceptor of the ciphertext does not have the shared key and cannot recover the original plaintext message sent by A. Since only A and B know the shared secret key, B can assume that the decrypted message corresponds to the original sent by A and that it has not been altered. As we shall discuss later in conjunction with message authentication, additional provisions may be needed to fully authenticate the message.

Commonly used symmetric-key algorithms are public and thus follow the principle of open design. They are usually variants of stream ciphers that operate one bit at a time and block ciphers that work on blocks. One of the most widely used is the Advanced Encryption Standard (AES) specified by the US National Institute of Standards Technology (NIST) [12] and adopted by many government agencies. AES operates on block sizes of 128 bits with a choice of key lengths: 128, 192, and 256 bits. Its predecessor DES (Data Encryption Standard) was criticized for weaknesses due to a shorter key length of 56 bits, and it has been largely abandoned after being broken with several publicly documented variants of brute-force attacks.

Key Exchange

Nodes wishing to use symmetric cryptography need to securely establish secret keys. One way to accomplish this is through a message exchange using public-key cryptography. However, that imposes the burden of implementing a public-key crypto system which may be too much overhead if it is only needed for the key exchange in an otherwise symmetric crypto setting. Diffie and Helman have described an algorithm for establishing of shared keys between two parties without the need to have any additional prior knowledge, such as of each other's public keys.

It basically works by having the two interested nodes jointly construct the key by incremental execution of a specified algorithm with the exchange of intermediate results and parameters over an insecure channel. The algorithm has the property that, by knowing each other's intermediate outputs, both nodes compute the same secret key. It also has the property that it is infeasible for a third party to compute the same secret key by knowing the chosen algorithm, parameters, and partial outputs exchanged by the two nodes. The key benefit of this approach is that the identical secret key is computed by both interested parties without having to actually exchange it. Once they have the key, the two endpoints can use it to encrypt and decrypt messages in a symmetric cryptographic scheme of their choice, such as the AES, with a mutually agreed-upon key length.

The math of these algorithms is somewhat involved and beyond the scope of this book. The original algorithm proposed by Diffie and Helman was based on the principles of modular arithmetic. More recent implementations use computations based on elliptic curves.

A problem with this algorithm is that it is susceptible to a man in the middle attack where a malicious third party intercepts the key computation messages and deceives the two parties by substituting a response of its own. In this way, an attacker can establish its own encryption keys with each of the two original endpoints and view and alter their message at will, while they believe that they are having a secure communication. This problem is due to the lack of prior authentication that may be accomplished using the mechanisms discussed in the subsequent sections.

Message Authentication

Presented encryption schemes provide various combinations of message confidentiality, integrity, authentication, and nonrepudiation (by means of digital signatures) that are summarized in Table 5.2. Rows one and two in Table 5.2 refer to the encryption of an outgoing message from sender A to B that A encrypts using B's public key or its own secret key, denoted as $E_{vB}(P)$ and $E_{sA}(P)$, respectively. As described in the previous section, these modes of encryption and decryption can be used to ensure message confidentiality or authentication, but not both. The double transformation depicted in row three can do both, but it is more computationally intensive. The last row refers to the symmetric encryption using a shared secret key.

For the purposes of this discussion, message integrity is defined as the assurance that the message has not been tampered with and altered in transmission. In other words, the message is received exactly as sent by its originator. Message authentication may be required to confirm the claimed identity of the sender.

Message authentication in rows two and four in Table 5.2 is labeled as partial, because in some cases the recipient B may not be able to completely authenticate the message upon decryption. This can happen if a forged ciphertext is provided to it by some adversary who claims A's identity and who does not even have to know the secret key. An imposter can generate a random bit string and present it as the ciphertext to B. If decryption of such message yields some plaintext, B can be

Table 5.2 Characteristics of cryptographic schemes

Cryptographic Scheme	Confidentiality	Integrity	Authentication	Digital signature
Asymmetric, $E_{vB}(P)$	Yes	Yes	No	No
Asymmetric, $E_{sA}(P)$	No	Yes	Partial, yes w MAC	Yes
Double transformation	Yes	Yes	Yes	Yes
Symmetric, shared key	Yes	Yes	Partial, yes w MAC	No

deceived into accepting the plaintext as a legitimate message sent by A. This problem can be somewhat ameliorated in exchanges where the original message has some known structure, such as the syntax that can be validated or prose in a known language. Many IoT systems use variants of JSON for message serialization whose syntax provides the structure that can be validated. Since the adversary is just creating random bit strings masquerading as ciphertext, its decrypted version may result in gibberish that does not pass the syntax verifier or correspond to the words in a language. In general, the structuring scheme does not work for binary transmissions. Credible message authentication may be provided by the cryptographic techniques discuss in the next section.

Message Authentication Code (MAC) and Secure Hash

Cryptographic message authentication is a secure way to ensure that the received message corresponds to its actual original, i.e., the plaintext sent by A. This is accomplished by creating a *message authentication code* (MAC). Authentication based on MAC requires that senders and receivers share the secret authentication key, preferably a different one than the shared key that may be used by the two nodes for symmetric cryptography.

MAC is computed by the sender node using a known function and the secret key to produce a relatively short fixed-length value that serves as the authenticator, sometimes also referred to as the tag. Once computed, the MAC is appended to the message. MAC-generation algorithms have the property that it is not possible to alter the message without affecting the authentication tag. Upon receipt of the message, B uses the shared secret authentication key to compute the MAC on the payload part of the message and compares it with the MAC that was received. If the two match, the receiver can be assured that the message was not tampered with and was sent by A who is the only other party that has the secret key used for MAC generation. For security of authentication, it should be computationally infeasible to compute a valid authenticator (tag) of the given message without knowledge of the key. To be of practical value, computation of the MAC itself should be relatively simple when the encryption key is known.

Note that the message itself need not be encrypted to be authenticated. MAC authenticates the sender and message integrity, but it does not provide message confidentiality. This may suffice for some types of IoT applications where data integrity may be more of a concern than confidentiality, such as reports of the non-sensitive sensor data. If confidentiality is also required, the message itself should be encrypted using some other symmetric or asymmetric cryptographic scheme. As usual, higher security means higher overhead with increased resource requirements and latency. Therefore, its level at the particular links and even for the specific types of exchanges should be balanced and chosen commensurate with the overall system requirements.

Hashing is a commonly used mechanism for creating unique values based on the content of a file. Hashing functions produce a signature of a fixed length that may be used for file identification. It is used in many applications, such as to verify file integrity or to detect identical files in storage management for deduplication. With the addition of a secret key to compute the hash, hashing can be made secure and computable only by the communicating parties who have the secret key. Secure hash functions are a popular choice for implementing message authentication algorithms.

The US National Institute of Standards Technology (NIST) has published several specifications for the Secure Hash Algorithm (SHA), known as the SHA series, that are evolving to address the weaknesses found in previous versions. The SHA-2 family, known as the SHA-256 and SHA-512, support different block sizes and use 32-bit and 64-bit words, respectively. SHA-3 algorithm uses the same hash lengths as the SHA-2 but differs in terms of its internal structure [13]. The Secure Hash Standard (SHS) was published separately [14].

Security Principles

Several decades ago, Saltzer and Schroeder [15] outlined general design principles for protection mechanisms in OSs. Many of them have stood the test of time and are followed as security design principles in contemporary distributed systems and on the Internet. They serve as useful guidelines for security designs in IoT systems, and their elements are instantiated in several hardware, software, and networking mechanisms discussed later in this chapter. The original principles slightly restated are:

- *Least privilege*. Every user and every process should use the least set of privileges necessary to complete the task.
- *Separation of privilege*. Where feasible, the protection mechanism should depend on satisfying more than one condition. A classic example is two keys held by different people required to unlock a bank vault.
- *Least common mechanism*. Minimize the use of mechanisms that are shared by multiple users. This is because shared mechanisms can lead to unintended leakage or interference.
- *Economy of mechanism*. A simpler design is easier to test and validate.
- *Complete mediation*. Every access request should be checked for authorization. It also requires that proposals to gain performance by remembering results of prior authority checks should be viewed with skepticism.
- *Fail-safe default*. Access rights should be acquired by explicit permission only, and the default should be the lack of access. Base access decisions on permission rather than exclusion.
- *Open design*. The design of the security mechanism should not be secret, and it should not depend on the ignorance of attackers. This implies the use of a cryptographic system where the algorithms are known, but the protection keys are secret.

- *User acceptability*. The mechanism should provide ease of use so that it is applied correctly and not circumvented by users.

Two additional principles listed in the paper are:

- *Work factor*. Stronger security measures should make attackers work harder, i.e., increase the difficulty of breaking into the point of making the effort not worthwhile.
- *Compromise recording*. The system should keep attack records to (help to) discover unauthorized use. Such records may be tampered with by successful intruders.

The more widely used principles are the least privilege, open design, fail-safe defaults, separation of privilege, and the least common mechanism. Complete mediation should be strived for, but it can be difficult to enforce in complex distributed systems. The economy of design may be hard to find in some of the today's complex security mechanisms. The work factor is useful advice for encouraging would-be intruders to try elsewhere. Recording of security events and compromises is part of the security monitoring discussed later in this chapter.

Endpoint Security

IoT system security requires implementation of some levels of protection in all components, including edge nodes, communication links, cloud, and authorization of user access. These design choices need to be coordinated and balanced to achieve the overall system goals as informed by the risk assessment and threat analysis.

In this section we describe a range of techniques and hardware and software assists commonly used for implementation of endpoint security. Depending on their roles and importance in the overall system, specific IoT endpoints may deploy some or combination of these mechanisms to achieve the desired level of security.

Basic objectives of IoT endpoint security include:

- Node identification and authentication.
- Secure state – bringing up and maintaining.
- Secure communications.
- Security monitoring and attestation.

Node identity uniquely differentiates an endpoint from all others and allows it to be referred to without ambiguity. It may be used for many purposes, with addressing and authentication being the more obvious ones. An endpoint identity needs to be unique at least in the domain and name space in which it operates. Globally unique identities are often preferred. They may be generated as such by the system or constructed from the domain-specific identities by prefixing them with the unique name of the domain. An endpoint may have one or multiple identities. For security purposes, it is desirable to be able to validate specific claims of identity. This may be accomplished by having the claimant use security credentials to support the claim. There are many variants of security credentials, including passwords, biometrics, or certificates. In security-conscious systems, a common approach is to use the cryptographic certificates that bind the public keys to the identity and are cryptographically signed by a trusted certificate authority (CA).

Other aspects of IoT endpoint security, including bringing up and maintaining node and communication security, are usually implemented with the assistance of hardware security modules and mechanisms which are described in the sections that follow. Network security is covered in a separate section.

Hardware Security Modules (HSM)

Use of hardware security modules (HSMs) is becoming a common way to strengthen and accelerate security functions in IoT systems. HSMs are hardware-based dedicated security components, usually consisting of a separate processor with its own private storage. They typically provide a safe isolated execution environment for implementation of security functions, usually coupled with the capability to generate and safeguard cryptographic keys that can be used for authentication, platform integrity, and secure communication. One of the important functions of HSM is to assist in bringing up of a node in a known good starting state.

Sections that follow describe several modalities of hardware security modules that may be used to harden IoT endpoint security, including Trusted Platform Module (TPM) and Trusted Execution Environments (TEEs).

Trusted Platform Module (TPM)

One of the most widely used implementation of HSM is the Trusted Platform Module (TPM), defined by the Trusted Computing Group [16] and now also an international standard ISO/IEC 118892015. The term TPM is somewhat overloaded in the sense of being used to refer to specifications as well as implementations. Earlier versions of the specifications provided hardware characteristics of the TPM, and the more recent are library specifications of functions that may be implemented in hardware or software [17]. The rest of this section describes some of its salient features relevant to IoT systems and representative of HSM functions.

TPM implementations are incorporated in millions of computer-based systems such as PCs, servers, and network routers. TPM also has specification profiles for other uses, such as automotive systems.

TPM specification may be implemented in a number of forms, the most common of which, in decreasing order of security, are (1) discrete TPM, a separate dedicated hardware component; (2) integrated TPM, into other processing elements; (3) firmware TPM in a separate protected execution environment; and (4) software TMP emulator. Systems that support virtualization may implement a virtual TPM that provides TPM functionality to each virtual machine.

Trusted Platform Module (TPM) technology is designed to provide hardware-based, security-related functions. A TPM chip is a secure crypto processor that is designed to carry out cryptographic operations. The chip includes multiple physical security mechanisms to make its security functions tamper resistant to malicious software. Some of the key components and capabilities of a TPM include:

- Protected capabilities – commands for exclusive access to shielded locations, storage of cryptographic keys, and registers used to authenticate the platform and report platform integrity measurements.
- Integrity measurement, logging, and reporting – to attest to integrity of the reported measurements.
- Their usage and operation – briefly described in the remainder of this section.

Root of Trust (RoT)

A common approach to securing the integrity of platform system software is by establishing a Root of Trust (RoT) for the initial system start-up and then subjecting each subsequent piece of code to verification of authenticity prior to its execution. In practice, RoT is some piece of code or hardware that has been hardened and either cannot be modified at all or requires cryptographic credentials to do so.

Most computer systems upon powering up transfer program control to a known address that contains executable code. The start-up code configures programmable control registers for the platform, including built-in input/output controllers, into a known initial condition. In PCs this is traditionally referred to as the Binary Input Output System (BIOS). At least the initial parts of BIOS are stored in a non-volatile memory that may also be read-only and thus unmodifiable. After completion, the BIOS transfers program control to another known location where the loader of the runtime system, typically an operating system (OS), is located. This process continues with loading and activation of the OS which in turn loads the other required components, such as libraries and network stacks, and ultimately the applications.

Securing of this process starts by executing an immutable trusted piece of code that is provided by the platform manufacturer. This is often referred to as the Core Root of Trust (CRoT). It may be embedded in the initial part of the BIOS or its more recent incarnation in the Unified Extensible Firmware Interface (UEFI). UEFI [18] provides basically the same function, but it can be programmed in a high-level language and provides the additional flexibility of interfacing with external modules. Those modules, known as option ROMs, are typically supplied by the manufacturers of the third-party I/O components that are included in the platform, such as the network and external storage controllers, to provide the necessary hardware initialization sequences.

Measured and Secure Boot

TPM provides assistance in the subsequent steps of the trusted boot process. It provides a secure way to characterize subsequent pieces of software before they are executed and to record the order of execution that it can attest to. TPM can compute secure digests of software modules prior to their execution by using its built-in crypto engines, such as the SHA. Digests may be computed on the software binary image and optionally combined with some other platform attributes to reflect a specific configuration. TCG and security nomenclature define this process as the *measurement* of software. TPM stores the results of measurements in one of its secure registers, called Platform Configuration Register (PCR). There is a number of those registers, some of them dedicated to recording specific measurements, such as the BIOS, option ROM, and Master Boot Record (MBR) for the OS boot. PCRs are reset to 0 after each TPM boot to ensure that the recorded measurements reflect the latest system power-up sequence.

In general, IoT nodes operate continuously with power cycling and rebooting being the exceptional rare events. However, protecting the system in all phases of its life cycle is important since security barriers are usually lowered during the boot time before the security agents, firewalls, and virus scanners are activated, which makes them comparatively more vulnerable to attacks and thus attractive to exploits. One of the forms of attack is to induce power glitches, thus causing repeated reboots and potential exposures.

In each power-up sequence, the TPM measures the successive pieces of software that are executed and records results in the corresponding PCR registers. This process is usually referred to as the *measured boot*. Some TPM implementations provide the capability to securely log and combine measurements, thus recording the order of all events in the current sequence and in prior boot sequences that may reveal tampering attempts.

Measuring the boot process does not guarantee integrity of the executed software. This is because the TPM or its enclosing platform cannot determine if the measurement corresponds to a known good state. For that to happen, the known good digests for each piece of the executed software would have to be provided and securely authenticated by a trusted third party, such as the software manufacturer with a valid cryptographic certificate. This approach is implemented in some platforms as the *secure boot*, where each measurement is compared and validated against the known good signature before being executed. If a mismatch is detected, the boot is halted, and the platform is prevented from entering an untrusted state. This ensures that a compromised platform will not join the system to inflict potential damage. On the down side, the platform is effectively rendered inoperable which usually requires manual intervention to restore or replace it – a process that does not scale well in installations with large number of nodes with different levels of trust, such as IoT.

Secure boot may be difficult to manage in IoT systems since it requires trust certificates and signatures up to date for the many components in a multi-vendor environment with frequent updates. When measured boot is used instead, the additional step of attestation is required to establish trust in the node.

Security has a transitive property, often described as "if A trusts B and B trusts C, then A trusts C." This can be applied to the secure or measured boot by creating a chain of trust where each validated component extends the chain of trust starting with the original Core Root of Trust. The process can be extended to a secure OS if one is used and even to applications if they are certified.

TPM aids in this process by securely authenticating the node and measurement values to an external attestation entity that evaluates the state by comparing supplied measurements to known good values. For this reason, TPM is usually said to be the Root of Trust for Measurement, as distinguished from the Core Root of Trust described earlier. The attestation authority usually resides in the IoT system, and it can manage software signatures from all legitimate and trusted pieces of software that are distributed to endpoints, including applications, microservices, and custom-developed code unique to a specific system. TPM also acts as a Root of Trust for Reporting by exposing the shielded measured values and by digitally signing those reports for verification of integrity by the attestation entity.

Data protection in rest and in motion is a potential security concern at all levels of an IoT system. As discussed in chapters "The Edge" and "Cloud", data can be stored at various points in the system, including edge nodes and the cloud. Data confidentiality in storage may be protected using encryption to make it unintelligible to all but the authorized applications. TPM may be used to provide private keys to encrypt storage and thus preserve its integrity. When applied in this manner, the TPM can act as the Root of Trust for Storage.

Secure Software Execution

Hardware TPM can facilitate secure bringing up of a node and possibly the application software, but it does not provide a secure execution thereafter. Securing of software at runtime is primarily based on the notion of isolation of execution environments and separation of trusted software from the untrusted one. This usually works better with some degree of hardware assistance.

Higher-end general-purpose processors have built-in mechanisms for protection of address spaces and levels of access privilege in the form of memory management hardware and guarded instructions for transitions between contexts, such as applications and OS address spaces. Separation between running processes is generally enforced by detecting and preventing attempts to address storage outside of their allocated memory space as defined by the page-map tables and segment limit registers. Unfortunately, these may be circumvented by adversaries who exploit weaknesses such as buffer and stack overflows.

Stricter separation may be provided by physically separating execution resources, such as processor and memory, for trusted and untrusted execution zones. Common methods for separation that could be used in IoT systems include secure elements,

Trusted Execution Environments, and virtualization. They are briefly described in the remainder of this section.

Secure Element (SE)

Secure elements are tamper-resistant subsystems capable of securely hosting code and confidential data, often coupled with a dedicated processor for execution of secure code. SEs are usually self-contained systems that may be physically integrated but operate separately from the platform's main processor and memory. They are often outward facing in the sense of providing specific services and authentication to external parties, such as merchants and mobile network operators. This makes them secure for their intended use, but not very suitable for more general security uses. With multiple applications being stored and their processes executed within a single device, it is essential to be able to host trusted applications and their associated credentials in a secure environment. Examples of this include authentication, identification, signatures, and PIN management, all of which are needed for different consumer services and require a protected environment to operate securely.

Smart cards and SIM (Subscriber Identification Module) cards were the early instantiation of secure elements. SEs are an evolution of the traditional chip that resides in smart cards, which have been adapted to suit the needs of an increasingly digitalized world, such as smart phones, tablets, set-top boxes, wearables, connected cars, and other Internet of Things (IoT) devices. More recent variants include Universal Integrated Circuit Card (UICC), embedded SE, and secure (micro) digital cards, microSD. UICCs are used as an evolution of SIM cards, as they support a larger variety of network types and additional applications that may be used by IoT nodes for services in the licensed spectrum, such as USIM to support Universal Mobile Telecommunication Systems (UMTS), High-Speed Packet Access (HSPA) networks, and Long-Term Evolution (LTE) networks. Embedded SEs (eSE) are tamper-proof chips that may be embedded in near-field communication (NFC) chips or directly into smart phones to support applications such as electronic wireless or contact payments.

Trusted Execution Environment (TEE)

Trusted Execution Environments (TEEs) are hardware-assisted segregated secure execution environments that were originally designed for processors commonly associated with mobile platforms. A TEE provides two important security properties: (1) secure storage of cryptographic keys and (2) dedicated storage and execution of security code. Some of its implementations also provide hardware accelerators for offloading and speeding up the execution of cryptographic operations.

Use of a TEE allows complete separation of trusted and untrusted portions of the system. Security applications and sensitive cryptographic assets may be completely sequestered in the TEE, while the regular, untrusted applications are executed with the main platform resources and a regular operating system (OS), sometimes referred to as the "normal" Rich Execution Environment (REE).

A processor block diagram with a TEE is depicted in Fig. 5.4. As indicated, a TEE usually consists of a separate processor with private secure storage for code and cryptographic secrets, crypto processors, or accelerators and sometimes even trusted peripherals. It usually runs a trusted OS and may provide additional functions for cryptographic operations and secure communications. The trusted kernel is securely booted upon each system start-up. The TEE may be used to provide system security functions and even to host a management agent that communicates securely with a remote security-management system authority. It may provide attestation, receive security reports, and refresh node's security policies and credentials as necessary.

A TEE may also provide isolation between trusted applications themselves and further secure the system by preventing regular interrupts and direct memory access (DMA) hardware to access its protected storage or to invoke any trusted actions.

The only way for the regular applications running in the REE environment to request TEE services is via the documented application programming interfaces (APIs). They provide a single controlled point of access with strictly defined parameters and controls. A trusted service that executes in TEE may require an invoking REE application to authenticate itself via credentials or login and may further

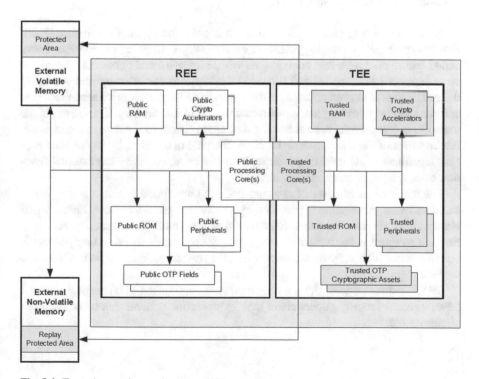

Fig. 5.4 Trusted execution environment [19]

restrict the type of services rendered based on the requester's identity and perceived level of trust.

In addition to REE APIs, a TEE typically provides several other classes of APIs. They usually include private APIs for use only by the trusted applications; user interface APIs with trusted peripherals for secure user authentication via PINs, passwords, or biometric means; and others such as secure sockets for communication.

There are several specifications and implementations of TEEs. Global Platform [19], an industry association with participation of major hardware and software vendors and users such as credit card companies, provides several publicly available specifications for SE, TEE, and Trusted Platform Services (TPS). ARM provides a TrustZone technology [20] which is an architectural specification very similar to the Global Platform, which was itself derived from an initial ARM proposal, and several modalities of hardware support for SE and TEE implementations. Several other vendors provide specialized security hardware compliant with various specifications. Open-TEE [21] is an open-source project for platform-independent implementation of virtual TEE. Google has an open-source project for a trusted TEE OS, called Trusty TEE, for Android systems [22].

Software Guard Extensions (SGX)

A variant of securing parts of an application is provided by Intel's Software Guard Extensions (SGX) technology [23]. It creates secure parts of an address space, called secure enclaves, that contain data whose integrity and confidentiality are to be safeguarded and specialized processor instructions for creating and accessing them. The idea is to protect sensitive data even from the higher privileged processes and the OS that normally has unrestricted access to all memory. Thus, even if the main OS is compromised, an attacker cannot gain access to the data in the enclave. Enclave memory is encrypted, and it cannot be read in the clear by any outside process regardless of its level of privilege. It cannot be accessed by the external function calls, jumps, or stack manipulation.

The mechanism consists of a secure enclave launch and provisioning by measured software. Attestation is provided by an external provider after which internal attestation is possible for the applications that have multiple interacting enclaves. Enclave memory is encrypted for safeguarding even when its enclosing process is inactive, and it can only be accessed when the trusted code decrypts it for its private use.

SGX illustrates one of the possible hardware approaches to increasing security. Several recent security papers claim that its protection scheme can be or has been circumvented.

Virtualization

Virtualization, among other things, provides the means for isolating execution environments by provisioning and executing them in different virtual machines. It is usually implementing by providing a virtualization software, called the *hypervisor*, that runs directly on hardware and below standard operating systems. A hypervisor virtualizes a hardware platform by trapping all system calls and access to hardware from the virtual machines (VMs) and mapping them as necessary to maintain the illusion that they are controlling the actual hardware. It provides all necessary memory mapping and arbitrates access to hardware to maintain separation and consistency across all VMs. Since the underlying hardware is actually shared by active VMs, the hypervisor usually allocates system resources in equitable and proportional manner and in accordance with their resource allocation. Many contemporary processors in the PC and server classes contain hardware assists for virtualization to speed up the execution of the hypervisor and to minimize the performance impact of the overhead that it introduces.

Form the security point of view, virtualization provides the benefits of isolating applications. It makes perfect sense to locate system security functions in a separate VM and thus protect it from the interference and tampering from other applications. Placing of other applications into separate VMs limits the impact of intrusions to only those that are directly affected and allows the rest of the system to operate relatively unscathed. With proper monitoring and privileges, a security system may detect an intrusion, shut down the affected VM, mitigate the problem if possible, and activate a known good replacement to restore the function.

Virtualization additionally provides an option for hardening the security of legacy systems, such as old SCADA installations running on outdated versions of OS with known weaknesses. In such systems it may be possible to install a hypervisor and run the legacy application unchanged in a VM with its version of the operating system installed as a guest OS. Then another VM may be created to host a security application, to attest the system and actively monitor the legacy VM for intrusions, and to execute security policies in effect to mitigate incidents should they occur.

For all this to work, the hypervisor itself should be trusted. It should be brought up via a trusted boot process and preferably attested before commencing operation. An example of how this can be accomplished is provided by Intel's Trusted Execution Technology (TXT) [24]. It extends the Root of Trust all the way to the hypervisor by means of measured launch process that starts by provisioning into the platform TPM known good values for the BIOS and hypervisor and measuring each before the execution. If attestation with stored PCR values for known good values matches the measures, the process completes with the activation of a trusted hypervisor presumably free of malware such as rootkits. If mismatches are detected, the software is assumed to be suspicious, and the TXT invokes the policy action that it is provisioned with. This usually means contacting an authorized external entity to report the problem and take the appropriate action.

TXT forms a trusted computing base (TCB) consisting of the processor, chipset, TPM, and flash that contains BIOS and TXT initialization code. In addition, the hypervisor code needs to be approved and its proper signature characterized by Intel and the software vendor. This works only for the TXT-enabled hypervisors, and the process needs to be repeated for each new release of the software.

Endpoint Hardware and Software Security

Figure 5.5 illustrates some common combinations and variants of platform software. It also includes some of the hardware security enhancements discussed in this chapter, such as the TPM and TEE.

As indicated, the execution environments in IoT nodes vary based on the needs and on the nature of hardware resources available for allocation. The simplest form is the embedded systems with some form of runtime to support application execution. These may include language libraries, bytecode interpreters for interpreted languages, just-in-time compilers, and possibly some form of IoT framework middleware. In such arrangements, referred to as the bare-metal configurations, applications may need to have awareness of the platform hardware, peripherals, and address map. Such systems usually do not include additional hardware provisions for security.

Addition of an operating system, a general-purpose one or a real-time OS (RTOS), provides a more structured environment for execution of multiple applications. The OS manages and allocates system resources to executing applications, and it may also provide some common functions, such as implementation of network stacks. Shared libraries and IoT frameworks may be included for additional

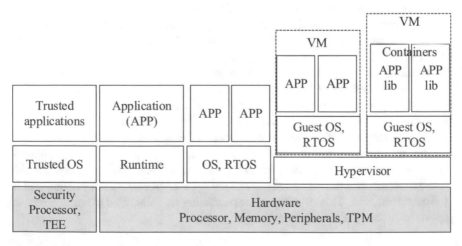

Fig. 5.5 IoT endpoint hardware and software combinations

support. All of those are usually made available to the executing applications via the API calls. Depending on the hardware capabilities, such platforms may provide different degrees of address-space isolation and separation between the running processes, but they have limitations and may be circumvented as described earlier. Security may be strengthened by the addition of hardware support for security, such as TEE and TPM.

Further enhancement of security may be accomplished in systems with virtualization which supports stronger separation and isolation between sets of applications. This usually works well for applications that are not tightly coupled in terms of common functionality and collaboration by sharing memory and direct transfers of control. As mentioned earlier, with a trusted hypervisor and virtualization, security breaches may be confined to the affected VM and potentially mitigated by actions of a protected security VM.

In systems with sufficient resources and proper OS support, containers may be used to further encapsulate application environments. *Containers* are essentially a packaging mechanism to bundle applications and their dependencies into a runtime system that may be dynamically placed onto a machine for execution. They can be executed on richer operating systems that support isolated execution environments, usually with their own namespaces and file access. A container contains an application and all its dependencies, such as the specific versions of the libraries required for its execution. Although not shown as a separate configuration in Fig. 5.5, containers do not require virtualization and can run directly on the supporting OS.

Containers simplify application deployment and portability by allowing a container to be installable on any supporting server or VM with sufficient memory, as opposed to limiting the placement choices only to the subset of servers that are an exact match for their runtime dependencies. Containers are widely used in cloud systems because of their ability to dynamically scale and migrate applications in response to the system changing load and needs. There are many tools for managing and orchestrating groups of containers that collectively implement complex systems. Containers are also a useful and popular mechanism in IoT system for creating and managing applications that may run in the cloud as well as on the capable edge nodes, such as the fog nodes and gateways, to optimize and adjust the function placement at runtime.

From the security point of view, use of containers can provide additional safeguards since operating systems usually manage containers as separate processes and largely independent environments with isolation and protection for address spaces that may extend to the file system.

Network Security

The previous section discussed some common ways used to secure IoT endpoints. An IoT system can be comprised of a large number of nodes connected by a variety of networks that are constantly communicating to move data to the processing points and to distribute the resulting actions and commands. Obviously, one of the key objectives of IoT security design is to provide the mechanisms and policies to ensure confidentiality, integrity, and authenticity of those communications.

As discussed in chapter "Communications" in conjunction with communications, IoT networks are implemented using a layered design. Consequently, communications may be secured at different layers, including link, network, transport, and application. Depending on their function and on the network topology, the perimeter of security provided by the lower layers may be limited to only the segments of the network where they operate. In principle, end-to-end secure message delivery is possible only at the transportation and application layers. The Internet Protocol Security (IPsec), often used for authentication and encryption in virtual private networks (VPNs), bends this rule a little by providing security associations that have a session-like effect on both TCP and UDP packets that it transfers.

Link-level security is valuable especially in the wireless network segments because they are easy to eavesdrop on. A good example of this is provided by the widely used variants of the Wi-Fi Protected Access (WPA), WPA2 and WPA3. They have enterprise versions that require use of an external authentication server and personal versions that use protected shared keys that are more practical for the home use. Like most standardized security protocols, they are evolving over time to keep up with the changes in technology and to address the weaknesses as they are uncovered.

Most communication security protocols require authentication of endpoints, followed by the use of encryption based on some of the cryptographic techniques described earlier in this chapter. We illustrate what they look like in the next section using the example of the Transport Layer Security protocol because it provides endpoint security and contains an elaborate version of the type of handshaking dance used for authentication and establishment of keys among the communicating nodes.

Transport Layer Security

Transport Layer Security (TLS) and Datagram Transport Layer Security (DTLS) are the most commonly used security protocols on the Internet, designed to work with TCP and UDP transports, respectively. TLS is a successor of the Secure Socket Layer (SSL) which has been largely deprecated due to its publicly documented weaknesses. Security protocols are revised to address discovered weaknesses, and they generally go through a series of releases. SSL protocol had three major releases, and the TLS is in its fourth [25]. In principle, the latest releases should be used, and software packages that implement network security on IoT nodes should be updated to reflect them whenever possible.

The basic underlying mechanism of secure communication and the TLS, in particular, is for the two communicating endpoints to engage in a message exchange that allows them to authenticate each other with digital certificates, to agree on the cryptographic schemes and their particulars, and to exchange the key material necessary to proceed to the securely encrypted transfers of application data.

The basic steps in TLS are:

- Endpoint authentication and secure secret key exchange using asymmetric cryptography.
- Confidential data exchange using symmetric cryptography with shared secret keys.
- Message authentication using secure hashing.

TLS has two primary components: (1) handshake protocol, which is used to negotiate cryptographic modes and parameters, authenticate communicating parties, and establish the shared key material, and (2) record protocol, which uses parameters established by the handshake to protect traffic between endpoints and divides traffic into a series of records, each of which is independently protected using traffic keys.

The handshake component has three phases illustrated in Fig. 5.6. It starts by the client sending a "Client Hello" message to the server and providing the list of supported SSL/TLS protocols, a random number combined with the current datetime, session ID, compression methods, and cipher suites, i.e., the cryptographic algorithms that are supported, including the client's preferred one.

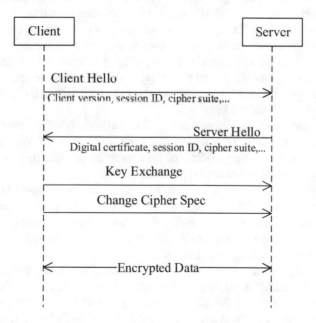

Fig. 5.6 TLS message exchange

For example, a version of the Windows 10 client using TLS 1.2 at this stage may supply a list of its supported combinations as follows:

TLS_ECDHE_ECDSA_WITH_AES_256_GCM_SHA384.

This particular string specifies using the TLS with the Elliptic Curve Diffie-Hellman Ephemeral (ECDHE) key exchange algorithm, the Elliptic Curve Digital Signature Algorithm (ECDSA) with the Advanced Encryption Standard (AES) symmetric key encryption and the key length of 256 bits in the Galois/Counter mode (AES_256_GCM), and the Secure Hash Algorithm with 384-bit key (SHA384) for message authentication. Ephemeral in ECDHE means that a new key is generated for each exchange. Session ID, if specified by the client, refers to a prior secure exchange, and the server may use it to retrieve the cached keys, if available, from a previous session between the two endpoints.

The server responds with the "Server Hello" message that includes the server-chosen level of security protocol which may not necessarily be the client's suggested preferred version; the server's choice of the cipher suite to use from the client supplied list; the server's random number combined with datetime, session ID, and compression methods to use; and the server's digital certificate (containing its identity and public encryption key). The server may optionally request the client to provide its digital certificate for verification.

The two random numbers are used to initiate and complete the key exchange using the agreed-upon protocol, such as the ECDHE. The two parties use the second phase of the message exchange to establish the sets of master keys that they will use for message encryption and MAC computations in each direction. The final phase of the handshake protocol is used for the two endpoints to synchronize a change to the chosen set of cipher suites and use them to encrypt the application data and to compute the MAC codes. Secure exchange of application data is then provided using the record protocol. It uses the parameters established in the handshake to divide up the payload into a series of records, each of which is independently protected using the established keys.

Some combinations of cipher suites in TLS provide what is referred to as the *perfect forward security*. That means that even if some secret keys are compromised, they cannot be used to decrypt previously exchanged messages that may have been intercepted and recorded by an adversary.

The Datagram Transport Layer Security (DTLS) secure transport layer protocol [26] is intentionally similar to TLS, but with some modifications to make it usable for datagram-based communication. These include sequence numbers and modification of the handshake protocol to include stateless cookie exchange for resistance against the denial of service attacks (DoS).

As illustrated, TLS and DTLS are quite elaborate systems, and their implementation can be demanding. A pragmatic IoT system designer may opt for incorporating an existing implementation from a well-known and trustworthy source and focus on selecting the cipher suite and key lengths suitable for the intended level of protection and capabilities of nodes on which it is to run.

Network Isolation and Segmentation

In IoT networks there are several distinct types of traffic that may have different security requirements. A coarse division may be made into the production (data plane) and the management traffic (control plane). Sensor data input is a production function that may have different security requirements than the actuation outputs. Since they can have impact on the physical world, actuation commands may be subject to the stronger confidentiality, integrity, authentication, and authorization controls. Management traffic includes asset status reporting, configuration updates, software updates, security status and monitoring reports, and commands with possibly different security requirements. It is a good design practice to separate different network data channels whenever possible. This may be accomplished by using a combination of techniques that include the use of separate physical networks, virtual private networks (VPNs), selection of different frequencies for different data channels on wireless networks, and using of separate TCP channels, ports, and sockets as well as different publish/subscribe topics for each.

In terms of physical links, IoT systems often include different segments of wireless and wired networks connecting nodes with different levels of protection and processing power. They may include private networks such as business networks, operations networks, and production control networks, as well as the public networks including the Internet. Accordingly, different network segments and assets connected to them, such as the endpoints and gateways, may have different levels of trust.

It is a good design practice to divide networks into segments that reflect physical connectivity and combine devices with the similar functional, security, and communications requirements. Such segments in effect constitute different trust zones in the system. Appropriate network components should be chosen and configured to protect the boundaries between the different segments and trust zones.

An illustration of a potential delineation of trust zones in an IoT system is provided by Fig. 5.7. Constrained devices and sensors with wireless networks may be the least trusted zone, followed by gateways and cloud systems. This is a general depiction of potential trust zones. In actual systems, fewer zones may be sufficient. For example, depending on their network positions and security protocols, nodes

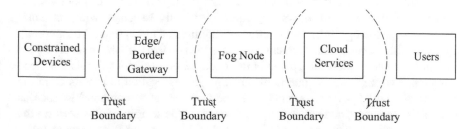

Fig. 5.7 Trust zones

in the fog segment may be included in the edge-to-cloud or in the cloud trust zone, as appropriate.

Each zone may have its own network access controls, node authentication rules, security policies, and protocols. At each boundary between zones, a mediating device such as gateway may act as a security guard by providing the necessary validation, filtering, and translation of messages.

Traditional methods for protecting network and segment boundaries in IT practice include mechanisms at different layers for filtering of link frames and packets. They can be performed by a variety of devices, including proxies and firewalls. A firewall limits the entry to a protected network by providing a single controlled point where the traffic can be inspected and suspicious content or sources denied. Traffic can be filtered based on network addresses, port numbers, and connection state that are sometimes referred to as the *stateful inspection*. Moreover, filtering may be performed at the transport and even application layers by analyzing messages and commands to allow only certain types of commands or users to cross the boundaries. This is sometimes referred to as the *deep packet inspection*.

In IoT systems, gateways are an obvious point for cross-boundary protection from the usually least trusted sensor and edge networks. They have direct connections to the potentially insecure legacy or constrained sensor networks on one end (the "south side") and the more mainstream IT-like networks with encryption and other forms of protection towards the cloud (the "north side"). Gateways usually have the power and resources to support the stronger forms of authentication and security protocols used in the core of an IoT system. This makes them a natural point to act as a bridge and security barrier between the edge and the core of an IoT system. Gateways can inspect and block or flag the traffic with improper authentication or from the untrusted sources and inspect the messages for conformance with the security policies and authorization levels, based on the nature of requests or actions that they may contain.

Security Monitoring

Security management and control functions in larger IT and IoT systems are often coalesced into a focal point to facilitate operations such as monitoring, definition, and distribution of security policies, credential management, secure boot attestation, and authentication of software updates. It is usually associated with a management console for operators to monitor and administer the system. An important function of the security management console is to visualize the system security state in real time and to alert the operators to abnormal or suspicious events that need attention. System security state is obtained as an aggregation to reports from the numerous security software agents deployed on the system components, such as endpoints, gateways, and networking equipment. Node security agents perform the local monitoring of status and activities related to the connections, communications, security events, and status changes. They send reports of their observations to the central control point via a secure channel and receive policy and security updates that need to be applied to the node on which they reside. The console is also used to create and administer security policies, generate status reports, and issue commands. It is common to host and execute the master control function at a relatively high level in the system hierarchy, such as the cloud.

Security monitoring, in the aggregate, has three important functions and components:

- Real-time state monitoring and analysis – keep track of the current system state, identify, and categorize anomalies and events that may indicate compromised devices and an attack in progress.
- Predictive monitoring and analysis of events – identify patterns that may indicate that an adversary has breached some parts of the system and is probing for weaknesses that may lead to future attacks.
- Forensic monitoring and analysis – record security data and events for event analysis, such as the compromised equipment and sequence of events that preceded attacks, successful and unsuccessful attempts.

The basic analysis can be performed by system operators based on their experience and observations of reported system state. Human analysis is often augmented and aided by the automated tools that are driven by prior knowledge and possibly artificial intelligence algorithms. Security analytics generally tends to be based on rule-based or behavioral analysis. *Rule-based* systems, also called signature-based, rely on a set of defined rules or known signatures of events to identify behaviors that are suspicious. When an event or a pattern that matches a rule is detected, it raises an alert to the operators and optionally triggers some automated response.

Behavioral analysis or anomaly detection systems primarily operate by learning the characteristics of good behaviors during the normal operation of a system. Following the training period, those systems are deployed to detect anomalous patterns that indicate or predict potential problems. Such systems may improve over time by incorporating characteristics of known bad behaviors learned from prior events and forensic analysis for faster detection and identifications of such patterns when they occur in the future. In practice, the two approaches are often combined because of their complementary strengths. When conditions requiring action are identified, some combination of operator advisory notifications and automated commands are undertaken to deal with the situation.

Forensic monitoring refers primarily to the logging of security observations and events. In addition to the local security logs, it is a good practice to also send log entries of at least major events to a remote location, typically the core of the security system. This facilitates construction of a global security state for a holistic view, as well as correlation and interpretation of events at the system level. Moreover, remote logging increases security by providing the visibility and persistence in cases where adversaries may gain control and delete security logs on the compromised nodes. Keeping security logs in secure storage of local nodes, where available, also helps to maintain their integrity.

A security monitoring system may also interface with the global malware monitoring services to get advanced notifications of major external activities and to update policies to proactively modify the security posture when heightened concerns are detected. In exchange, the system may also contribute its own observations of unusual events and attack attempts to the regional and global monitoring entities.

Security Incident Handling

While security monitoring helps to identify threats that are developing, an important part of the security management is to have a plan of action for dealing with incidents and breaches when they occur. As indicated, detection of developing threats should provide guidance to increase the level of perceived security threat and to activate the corresponding countermeasures and security policy updates. When incidents do occur, actions and procedures outlined in the incident response plan should be put into effect. This generally involves identification of the affected parts of the system and enhancement of security policies in reachable nodes and confinement through disconnection of affected components or related parts of the network. Depending on the nature of the breach, services to the affected nodes may be temporarily suspended and their messages and attempts to communicate blocked.

After an incident, recovery actions and procedures should be followed to restore the normal operating state of the system as soon as it is safe to do so. Forensic analysis of relevant logs should be performed to determine the root cause and the sequence of events that led to the attack to provide mitigation that would prevent similar attacks in the future.

Management

In addition to being secured, IoT nodes need to be managed in functional states and to operate with up-to-date versions of firmware and software. This is the task of the management subsystem. In addition, it is in charge of several operations in the life cycle of an IoT node that precede its production operational stage. In general, key stages of an IoT node life cycle may be summarized as follows:

- Manufacturing.
- Installation and activation (commissioning and provisioning).
- Operation.
- Decommissioning.

In the manufacturing stage, a node can be provisioned with the unique identity assigned by the manufacturer. It may be immutable and persist throughout node's lifetime. The node may also receive some security credentials that can be used to assist its automated "low touch" provisioning and initial authentication to the IoT system during the process of activation. The manufacturer may also add some machine-readable metadata, such as the serial number, device model and type, and possibly some functional description. The metadata format for these entries is not standardized, and, when used, it tends to be manufacturer specific.

The next major life cycle stage is when the node is installed and prepared to join the IoT system. This stage is also referred to as the commissioning and provisioning without a commonly accepted definition or clear distinction. It basically prepares a node for activation. Installation may be performed by the system integrator, and it includes physical placement of the node in its intended location, connection of external devices such as sensors, and connection to the network and to a power supply.

The next step, node activation, is commonly performed by the IoT system owner or its operator. The process is usually initiated from the backend sometime after the node installation. One way of establishing credibility of the claim of node legitimacy, often used in the installation and commissioning phases, is to give evidence of physical proximity to the device, such as reading the QR code with a phone camera or establishing near-field communication using Bluetooth or NFC-enabled devices. Node activation may be done individually or in bulk for large installations.

System management activities related to node activation typically include:

- Node ID assignment.
- Assignment of security credentials.
- Designation of system attach points.
- Registration.
- Updates.

The system assigns a node identity which is used for management purposes and as an anchor for identifying its data sourcing and actuation endpoints. It may be derived from or related in some way to the manufacturer-assigned identity. The node identity format may be humanly readable, for the benefit of system operators and debugging purposes, or just a unique machine designation as a string of letters and numbers. Some systems use both for efficiency and convenience by use of humans and applications and correlate the two to avoid ambiguity. The node ID identity needs to be unique within a system. For cross-domain data aggregations and big-data applications, it is often useful to have globally unique identities. This can be accomplished by prefixing node names by a unique registered domain name in which they operate or by using globally unique identifiers (GUIDs).

Security credentials are assigned to the node for authentication purposes. They may include a signed digital certificate and some additional cryptographic material, such as elements of keys to be used for encryption and secure communications.

Designation of system attach points generally includes items such as assignment of the home node to use when interacting with the rest of the system, brokers to use for publish-subscribe, or addresses to which to post the data. These are determined based on the node physical and network location, trust zone, and its capabilities and system role. For example, wireless sensors and smart things may be pointed to an edge router or a gateway and gateways to their corresponding secured cloud attach points. Commercial implementations of cloud endpoints are discussed in chapter "IoT Platforms" on platforms.

Registration of the node in the relevant backend systems and databases is necessary to recognize and authorize it for operational and management purposes. These systems may include IoT device registry, asset management systems, billing systems, ERP and CRM systems, and even interactions with the external systems to complete the provisioning, such as the mobile operator to activate device SIM cards.

System software and applications are usually loaded on the new nodes prior to or during their installation. Node application software includes data-processing functions, as well as management and security agents. Nodes are then provisioned with their security credentials and registered, after which they can join the system. Upon connecting, they may receive the latest updates of the firmware and system and application software if their provisioned versions are lower than the ones currently used in the system. Following the final updates, including security and management policies in effect, a node can start its normal operation.

Combinations of preparation, installation, and activation actions are referred to various systems as commissioning, provisioning, bootstrapping, and onboarding.

Once activated, the node commences its normal operation and becomes a productive member of an IoT system. During its operational phase, the management system monitors the states of active nodes with the assistance of the local management agents that report node status and assist in reconfiguration and firmware and software updates as instructed. Abnormal states and failures are flagged, and they may result in visual indications on the management console, alarms, or operator notifications. In general, given the potentially large number of nodes in an IoT system, management operations are automated, and the operators tend to interact directly only with the most important nodes, such as those at the key aggregation or processing points in the cloud.

Management monitoring and control operates in parallel with the security-focused discussed in previous sections. Depending on the implementation, security and management systems and node agents may be combined or can operate separately. The latter may be the case when independent commercial providers or open-source systems are used to implement the two systems.

The final stage in a node's life cycle is decommissioning, when it is removed from the system. At this point, it should have all its credentials and owner-assigned identity removed and be removed from or marked as decommissioned in system registries. This is necessary to prevent the node from connecting or revealing stored data if it is reused in another system. The intent is to protect integrity in commercial systems and to safeguard privacy and security in consumer uses that may be jeopardized if the device is sold or given away without being reset to the factory settings.

Privacy

Data privacy is a very interesting and sometimes fascinating topic, but its detailed treatment is beyond our present focus. In the remainder of this section, we briefly highlight some of its unique aspects related to IoT systems.

Privacy issues can be exacerbated in IoT systems that can directly interact with the physical world for two major reasons: (1) breaches can endanger and harm people and property, and (2) monitoring of the physical world can reveal sensitive information by direct observation or by inference in ways that users and privacy regulators may not be aware of. This is rather obvious with, for example, voice-activated AI assistants that listen to and may record every sound and utterance in a household, in a car, and almost anywhere when used with phones.

Even some seemingly innocuous IoT applications, such as the home energy monitoring that optimizes energy consumption by monitoring power usage and room occupancy, can reveal the times of presence or absence of occupants and detailed information on their activities in real time. These can include time-annotated record of use of electrical devices indicating activities such cooking, washing, listening to music, using a computer, printing, and sleeping. This can be accomplished by monitoring the power consumption to detect changes, such as spikes and drops when things are turned on and off, and by using the known device electrical signatures and patterns to disambiguate the events and to attribute them to specific devices. The results can range from a mere inconvenience to unwanted surveillance and even to the endangerment of people well-being and safety.

Protecting privacy is, or should be, a strong requirement in computer systems that collect and operate on user data, and even more so in those that contain IoT components, such as sensors and smart things. User data should be collected only with the subject's explicit consent, and they should be given at least some control and ownership of their data. This may include carefully designing and controlling data access authorization, and providing awareness to users, and possibly legal authorities, on items such as:

- Which personal data are collected?
- Who has access to data?
- Are data secured in transfer and in storage?
- How long are data kept? (forever or with a built-in expiration date).
- What are the notification procedures when data breaches are detected?

Some countries regulate data collection and privacy. The EU General Data Protection Regulation (GDPR) document [27] stipulates that data subjects, whose data are collected, have the right to access their data, the right to be forgotten, and the right to be notified on any breaches.

Protection of data privacy may collide with some industry practices and business models that are built on the, often opaque, collection of user data in exchange for the use of "free" services, such as social networks, Internet searches, location services, and mapping applications.

Data integrity rather than privacy is perceived as being the primary concern in industrial OT systems that collect production data owned by the corporations that they do not want divulged to the third parties. As those systems are being integrated with the IoT systems, they may need to address the privacy concerns as well.

Putting It All Together

In this chapter we discussed a number of security threats and vulnerabilities that Internet in general and IoT systems in particular may be exposed to. We also presented a number of security countermeasures that are generally known and used to mitigate those risks. In practice, attitudes range from the legacy OT myths such as "this could not possibly happen in our system" to the extreme concerns voiced by the security researchers that anything known even as a theoretical threat in any setting must be guarded against.

Design of the security plane requires a systems approach. The rational systematic approach to designing security in an IoT system is to start with a careful evaluation of risk and threats to the specific system at hand and to deploy a selection of security techniques and procedures that are both balanced and appropriate to the specific installation and conditions in which it operates.

Risk assessment and threat modeling steps identify what needs to be guarded against. They guide selection of security techniques that are appropriate for the particular components and segments of the system, including components and networking at the perception layer (sensors and things), gateways and fog nodes, WAN connectivity, and cloud services. Components at different levels usually have resource and operational constraints that limit levels of security that they can support and consequently determine their level of trust. The system should be designed to reflect that fact with the corresponding definition of trust zones with segmentation and isolation using the proper barriers between them. The chosen target level of security in each zone should guide the choice of hardware capabilities and security assists, types of cryptographic systems, and their strength determined by chosen key lengths. As indicated, security is primarily based on open design and secrecy of the keys [29]. Consequently, widely used operating systems and runtime environments provide implementations of security and a choice of cryptographic systems. IoT system designers do not have to and probably should not implement security algorithms from scratch – they just need to choose and configure available components to meet their design objectives.

Security design also needs to include procedures and mechanisms for provisioning node credentials and security material, such as the key exchange in symmetric cryptographic systems, and issuing of digital certificates and protection of private keys in the public-key systems. In addition, prudent IoT configurations should deploy defensive measures appropriate for the system at hand. Nodes should require certificate-based access authentication and disable or disallow password logins. They should close communication ports by default and keep open and monitored only those that are required for the operation of the specific protocols used by the IoT system. Remote and even local shell access should be disabled or completely removed when configuring node operating systems. Communication should be allowed only with the known set of trusted parties and services that may be white listed so as to exclude all others. Moreover, nodes should be directed to communicate only with a limited population of known entities that can be identified and with whom trust can be established through authentication. They are not, and should not be treated as, the general-purpose devices on the Internet that can be exposed to malware by downloading the code and scripts from almost any unverified source.

Physical security should also be taken into consideration, as some IoT devices may reside in facilities with controlled physical entry and sometimes actively monitored by security personnel. None of these provide security, but they may provide a reasonable basis for narrowing the list of exposures in the threat analysis and focusing on the most relevant and likely ones that remain.

Node life cycle management, provisioning, and system admission are largely the province of the control plane – security and management system. They are also responsible for distributing updates and maintaining nodes in the best known working shape in terms of firmware, software, and security policies. Both systems operate by constant monitoring of nodes in operation to detect anomalies, such as failures, security attacks, and breaches. Monitoring of intrusion attempts allows the system to dynamically adapt its security posture to reflect changing conditions and to execute the defined procedures and containment of affected system segments in case of breaches.

As a reminder and a potentially useful checklist, Microsoft researchers [28] have summarized their view of security properties for the highly secure (IoT) devices. Specifically, seven properties are identified as listed in Table 5.3.

They include many of the techniques and mechanisms described in this chapter and echo some recommendations advocated by the security principles described earlier. Defense in depth is a design principle that states that multiple layers of defense should be provided throughout the security system. The idea is to place multiple obstacles on the way of intruders and to confine breaches to the isolated parts of the system. This can, at the very least, delay the advance of the attack, thus allowing more time for some defensive measures to be taken. These measures may include a combination of technical, physical, and administrative controls.

Table 5.3 Properties of highly secure devices

Property	Description
Hardware-based root of trust	Unforgeable cryptographic keys generated and protected by hardware. Physical countermeasures to resist side-channel attacks
Small trusted computing base	Private keys stored in a hardware-protected vault, inaccessible to software. Division of software into self-protecting layers
Defense in depth	Multiple mitigations applied against each threat. Countermeasures mitigate the consequences of a successful attack on any surface
Compartmentalization	Hardware-enforced barriers between software components prevent a breach in one from propagating to the others
Certificate-based authentication	Signed certificate, proven unforgeable by a cryptographic key, proved the device identity and authenticity
Renewable security	Renewal brings device forward to a secure state and revokes compromised assets for known vulnerabilities or security breaches
Failure reporting	A software failure, such as a buffer overrun induced by an attacker probing security, is reported to a cloud-based failure analysis system

References

1. Bumiller, E., Shanker, T. (2012) 'Panetta warns of dire threat of cyberattack in US', *The New York Times*, 12 Oct [Online] Available at: https://www.nytimes.com/2012/10/12/world/panetta-warns-of-dire-threat-of-cyberattack.html (Accessed Dec 15, 2019)
2. Garcia-Morchon, O., Kumar, S., Sethi, M. (2019) 'Internet of things (IoT) security: state of the art and challenges, IETF RFC 8576, [Online] Available at: https://tools.ietf.org/pdf/rfc8576.pdf (Accessed Dec 15, 2019)
3. Schrecker, S. et al, (2016) 'Industrial internet of things volume G4: security framework', [Online] Available at: https://www.iiconsortium.org/pdf/IIC_PUB_G4_V1.00_PB-3.pdf (Accessed Dec 15, 2019)
4. Kushner, D. (2013) 'The real story of stuxnet', *IEEE Spectrum*, vol. 50, no. 3, pp. 48–53.
5. Musil, S. (2012) 'Crippling Stuxnet virus infected Chevron's network too', CNET, 8 Nov, [Online] Available at: https://www.cnet.com/news/crippling-stuxnet-virus-infected-chevrons-network-too/ (Accessed Dec 15, 2019)
6. SIMATIC WinnCC – System Overview [Online] Available at: https://new.siemens.com/uk/en/products/automation/hmi/wincc-unified.html (Accessed Dec 15, 2019)
7. Antonakis, M. et al. (2017) 'Understanding the Mirai botnet', *26th Usenix Symposium*, Vancouver BC, Canada [Online] Available at: https://www.usenix.org/system/files/conference/usenixsecurity17/sec17-antonakakis.pdf (Accessed Dec 15, 2019)
8. Microsoft (2018) 'Internet of things (IoT) security architecture' [Online] Available at: https://docs.microsoft.com/en-us/azure/iot-fundamentals/iot-security-architecture (Accessed Dec 15, 2019)
9. Diffie, D., Hellman, M. (1976) 'New directions in cryptography', *IEEE Transactions on Information Theory*, 22 (6), p 644–654.
10. Rivest, R. L., Shamir, A., Adleman, L. (1978) 'A method for obtaining digital signatures and public-key cryptosystems', *Communications of the ACM*, 21(2), p 120–126.
11. Cooper, D. et al. (2008) 'Internet X.509 public key infrastructure certificate and certificate revocation list (CRL) profile' IETF RFC 5280 [Online] Available at: https://tools.ietf.org/pdf/rfc5280.pdf (Accessed Dec 15, 2019)
12. National Institute of Standards (2001) 'Advanced encryption standard (AES)' FIPS 197, [Online] Available at: https://nvlpubs.nist.gov/nistpubs/FIPS/NIST.FIPS.197.pdf (Accessed Dec 15, 2019)

13. National Institute of Standards (2015) 'SHA-3 standard: permutation-based hash and extendable-output functions' FIPS 202, [Online] Available at: https://nvlpubs.nist.gov/nist-pubs/FIPS/NIST.FIPS.202.pdf (Accessed Dec 15, 2019)
14. National Institute of Standards (2015) 'Secure hash standard (SHS)' FIPS 180–4, [Online] Available at: https://www.nist.gov/publications/secure-hash-standard (Accessed Dec 15, 2019)
15. Saltzer, J. H., Schroeder, M. D. (1975) 'The protection of information in computer systems', *Proceedings of the IEEE*, 63(9), p 1278–1308.
16. Trusted Computing Group (2007) 'Architecture overview' [Online] Available at: https://trustedcomputinggroup.org/wp-content/uploads/TCG_1_4_Architecture_Overview.pdf (Accessed Dec 15, 2019)
17. Trusted Computing Group (2016) 'Trusted platform module library part1: architecture' [Online] Available at: https://trustedcomputinggroup.org/wp-content/uploads/TPM-Rev-2.0-Part-1-Architecture-01.38.pdf (Accessed Dec 15, 2019)
18. Unified extensible firmware interface (UEFI) specification version 2.8 (2019) [Online] Available at: https://uefi.org/sites/default/files/resources/UEFI_Spec_2_8_final.pdf (Accessed Dec 15, 2019)
19. Global Platform (2018) 'TEE system architecture v1.2' [Online] Available at: https://globalplatform.org/specs-library/tee-system-architecture-v1-2/ (Accessed Dec 15, 2019)
20. ARM (2017) 'Trust zone technology for the ARMv8-M architecture' [Online] Available at: https://static.docs.arm.com/100690/0200/armv8m_trustzone_technology_100690_0200.pdf (Accessed Dec 15, 2019)
21. Open TEE [Online] Available at: https://open-tee.github.io/ (Accessed Dec 15, 2019)
22. Trusty TEE [Online] Available at: https://source.android.com/security/trusty (Accessed Dec 15, 2019)
23. Anati, I. et al. (2013) 'Innovative technology for CPU based attestation and sealing' [Online] Available at: https://software.intel.com/sites/default/files/article/413939/hasp-2013-innovative-technology-for-attestation-and-sealing.pdf (Accessed Dec 15, 2019)
24. Greene, J., 'Intel trusted execution technology' [Online] Available at: https://www.intel.com/content/dam/www/public/us/en/documents/white-papers/trusted-execution-technology-security-paper.pdf (Accessed Dec 15, 2019)
25. Rescorla, E. (2018) 'The transport layer security (TLS) protocol version 1.3', *IETF RFC 8446*, [Online] Available at: https://tools.ietf.org/pdf/rfc8446.pdf (Accessed Dec 15, 2019)
26. Tschofenig, H., Fossati, T. (2016) 'Transport layer security (TLS)/ datagram transport layer security (DTLS) profiles for the internet of things' [Online] Available at: https://tools.ietf.org/pdf/rfc7925.pdf (Accessed Dec 15, 2019)
27. EU Data Protection Rules (2018) [Online] Available at: https://ec.europa.eu/commission/priorities/justice-and-fundamental-rights/data-protection/2018-reform-eu-data-protection-rules_en (Accessed Dec 15, 2019)
28. Hunt, G., Letey, G., Nightingale, E. B. (2018) 'The seven properties of highly secure devices', [Online] Available at: https://www.microsoft.com/en-us/research/wp-content/uploads/2017/03/SevenPropertiesofHighlySecureDevices.pdf (Accessed Dec 15, 2019)
29. Stallings, W. (1995) *Cryptography and network security*, Prentice-Hall New Jersey.

Chapter 6
IoT Data Models and Metadata

One of the major IoT system design challenges is to address the need for interoperability. As mentioned in chapter "Introduction and Overview", achieving interoperability is estimated to increase the potential IoT market value by 40% [1]. The usefulness of IoT systems generally increases with their scale, volume, and variety of data. Many complex IoT systems – for instance, smart cities – require coordination of different subsystems that may and often use different data formats. The target is to achieve semantic or, as the Industrial Internet Consortium referred to it [2], conceptual interoperability, i.e., "represent information in a form whose meaning is independent of the application generating or using it."

In IoT systems, the bulk of data exchanged between the machines and services represent sensor readings and actuator commands. These are typically device- and domain-specific numbers, e.g., the temperature readings, with annotations such as the nature of measurements and units of measure, and control strings that do things like change a set point. Applications (not humans) at the receiving end need to be able to "understand" and interpret the message data in order to process and to act on it appropriately. In other words, the two communicating IoT endpoints need to achieve semantic interoperability to interact properly.

As pointed out earlier, the Internet does not have the problem or the requirement to establish machine-level semantic interoperability. Textual information on the Internet is generally not interpreted by the receiving applications, such as browsers. They just use the embedded annotation and formatting instructions, such as HTML, to render the content properly. The content is mostly destined for human consumption, with the semantic definitions provided by the language in which the text is presented.

As it happens, IoT systems do have the problem and the requirement to achieve semantic interoperability. This implies formatting and annotating the data in a manner that is interpreted identically at both ends of a message exchange. The problem is that there is no commonly accepted definition or standard for representing data and semantics in IoT systems. In its absence, it is not possible to independently develop IoT endpoints that interoperate by virtue of adhering to the same standards

© Springer Nature Switzerland AG 2020
M. Milenkovic, *Internet of Things: Concepts and System Design*,
https://doi.org/10.1007/978-3-030-41346-0_6

and specifications, which was a key catalyst in the explosive growth of the Internet. In this chapter we highlight the nature of the problem and outline the commonalities in approaches for dealing with it. In the next chapter, we briefly review several IoT standards and proposals under development that tackle the issues of data representation and interoperability.

IoT Data and Information Models

IoT systems essentially interface the Internet to the physical world via endpoints directly connected to things that exist in and can interact with it. Edge nodes, such as sensors and smart things, have physical properties and states. For brevity, we will refer to them collectively as generic (IoT) things. They need to be modeled and represented in the digital form for processing and communication. The model of a thing needs to express and reflect its salient physical characteristics. An IoT thing description is an abstraction of a physical entity that participates in an IoT system software layer exchanges, e.g., to report status and initiate actuation where applicable.

Common IoT data are sensor readings with values and units of measure. In order to be useful to applications and services, IoT data need to include the identification of the types of physical entities that produced them, such as temperature sensors, units of measure (°C), and metadata defining conditions and context in which the data are captured.

For two IoT endpoints to interoperate, they must have a shared understanding and interpretation of the thing data semantics. In other words, all communicating parties need to use the same conceptual model of things that exist in their domain. One way to accomplish this is by having all parties use the same specification of an IoT information model.

An *IoT information model* is an abstract, formal representation of IoT thing types that often include their properties, relationships, and the operations that can be performed on them.

More specifically, a thing information model needs to specify what the thing is what it can do, thing properties and their values, and ways to interact with it. IoT information models and data representations typically include:

- Designation of the type of the physical thing, what it is and what it does, e.g., temperature sensor
- Data formats, how to represent and interpret it, e.g., values (string, number, int or float), engineering units if applicable
- Interactions, access methods – how to access the thing to obtain its state and data or activate intrinsic functions and actuate outputs

    Optional but quite common components of IoT information models also include:

- Metadata describing thing attributes and context in which data are captured

- Links to other objects, used to depict structural connections and for object compositing

Interoperability Using a Shared Information Model

Figure 6.1 illustrates two endpoints in an IoT system that intend to engage in data and control exchanges. As indicated earlier, all communicating parties need to use the same conceptual model of things that exist in their domain. A common way to accomplish this is by having all parties use the same specification of an IoT information model.

Software objects that represent things are stored on an IoT server, a software module that provides interface and representation of the thing, its state, and its behaviors. Figure 6.1 illustrates an IoT server representing the associated physical endpoint, designated as the thing. The IoT server internally implements the thing description as a cyber-world touchpoint for each particular instance using the format of the information model defined by the shared specification. An IoT server typically includes device drivers to access the thing and convert its state and physical-world measurement into digital representation. It also needs to implement the actions associated with the device when requested by the authorized parties via the specified access points. Depending on the system configuration and hardware capability, an IoT server may physically reside on the thing or on a proxy device, such as a directly attached gateway.

Following the Internet principles, clients and servers exchange payloads and rely on an information-model specification for proper encoding and interpretation of the

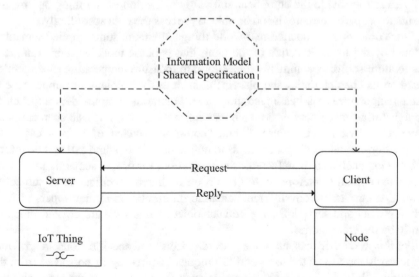

Fig. 6.1 Shared specification of information model

exchanged messages. The two endpoints are otherwise decoupled; they may be developed independently of each other and operate on different platforms and run-time environments. In this context, endpoint refers to software agents that are acquiring (producing) or processing (consuming) real-world data.

IoT clients are consumers of thing/server data and generators of actuation commands. IoT clients are commonly software agents associated with or acting on behalf of services and applications. A client can access server thing descriptions that it is authorized to and use them to issue queries and optionally command messages. The client uses the shared information-model definition to properly format requests and interpret responses that it receives. IoT clients may reside in the cloud or at some intermediate point in the system hierarchy, such as a peer thing node, gateway, or a fog node.

After having identified a thing of interest (through system directory or configuration file), the client typically consults the thing description on the server for a specification of how to access and query the target instance by exchanging messages using a mutually agreed protocol, such as HTTP or CoAP. After that, it can issue requests for data or command execution using actions defined in the thing description. The data-serialization method defines what the exchanged bits look like on the wire.

Data Semantics

One of the primary tasks of an IoT information model is to facilitate semantic interoperability, i.e., a shared understanding of what the data means. It is used by the IoT server to provide the compliant software abstraction of the thing and by IoT clients to properly interpret it and be able to process the data accordingly.

Since objects are modeling real-world things, there are some implied semantics defined by the intrinsic nature of the thing that is being modeled. For example, a temperature sensor is commonly understood to measure temperature of something based on its characteristics and placement, such as the ambient air temperature or water temperature in a heating/cooling pipe. For more complex devices, such as manufacturing machinery or even a home thermostat, their intrinsic characteristics may not be obvious in the sense of being "common knowledge." In those cases, the information-model specification needs to define or refer to some external definition of behavior, such as a manufacturer's specification, for proper semantic interpretation. Conceptually, interoperability based on a shared specification is subject to errors that can arise from different semantic interpretation by developers of individual clients and servers for a particular object when they are created independently by different parties.

For this and other reasons, some research efforts and specifications are gravitating towards the use of the tools and techniques inspired by or borrowed from the semantic web [3, 4]. They include ontologies to formally specify the semantics and behaviors of the defined objects. Machine-understandable descriptions often rely on the semantic web standards and machinery, such as the Resource Definition

Frameworks (RDF) [5] and the Web Ontology Language (OWL) [6]. In theory, this approach can provide a complete behavioral specification that is machine understandable and thus avoids ambiguity. When implemented, it can be applied at run-time for clients to obtain semantic description of objects that they encounter.

Conceptually, this implies the ability to discover a reachable object type for which the application has no pre-defined handler but can obtain its model from a known location or from the one pointed to by the object. This is somewhat easier to achieve when coupled with an information model that has a specification for the general class of devices and relies on the discovered model for details of specific features and instructions on how to use them. In general, this capability requires the definition and use of object taxonomies and ontologies.

General use of ontologies in IoT systems is some time away, as they have proven to be difficult to construct for physical entities or to grow beyond the illustrative local definitions into reaching a broad-based international consensus [7]. In the current practice, the tendency is towards developing IoT specifications and standards that include information models and may incorporate some elements of ontologies and machine-readable semantic descriptions in their thing descriptions.

Structure of IoT Information Model

There are many standardization efforts underway to define IoT information models, several of which are presented in detail in the next chapter on Standards. Individual specifications tend to cover different application domains – such as the industrial, automotive, and consumer – and differ in the specifics, terminology, and scope. However, they tend to follow the same basic modeling structure that is outlined below.

Thing software representations are often referred to as IoT information models and represented as (software) objects. Software objects that represent and model IoT things are often structured in ways that follow the common practice and terminology in object-oriented (OO) programming. They are also referred to by other names, such as thing descriptions and smart objects.

Use of the object-oriented representation is almost universal in the current definitions of information models. This is somewhat of a mixed blessing. On the positive side, the OO approach is familiar to many developers, and its basic structure of data, property, and method encapsulation is a fairly good fit for IoT information modeling. On the negative side, some of the common concepts and techniques in OO systems, such as class hierarchies, sub-classing, and class inheritance, do not always map well to the IoT system physical nature and structures. This sometimes creates the problem of the tool idiosyncrasies influencing the structure of the model, rather than having the model be an isomorphic reflection of the physical thing that it is intended to represent.

Table 6.1 illustrates a generic form of IoT information model representations. Details of its components are described in the sections that follow. Their primary purpose is to facilitate interaction and semantic interpretation of the exchanged data.

Table 6.1 Structure of IoT information model

Element	Description
Object type	Physical thing being modeled (endpoint, device)
Properties, attributes	Object attributes, data, metadata
Interactions	Ways to interact with object, access points, actions
Links	To other objects, compositions, and collections

Object Type

Within each specification, names of thing abstractions indicate the type of the physical thing or the phenomenon that is being modeled. This field is labeled as object type in Table 6.1 and is often modeled in object-oriented systems as a class. Object and class types directly correspond to the specific real-world things that are modeled by the specification. They can include a variety of sensors and devices, such as temperature, humidity or speed, door lock, air-handling unit, pump rotation, flow, valve, voltage, current, metered units (kWh, gals, or liters), microphone, thermostat, security camera, and refrigerator. It is customary to use a single object-class definition representing a category of physical devices or things, such as a temperature sensor, and to create specific object instances of that class for each individual physical instantiation of a thing in a particular IoT system.

Individual instances of things are assigned unique instance names to differentiate them. Instance names are used to uniquely identify each corresponding physical endpoint in a system. For example, all instances of temperature sensors in a given specification will share the common object/class type name within that specification, say "temperature sensor," with individual sensors being assigned instance names, such as "ts123" or "bedroom1". These names are often defined within assigned name spaces and are unique within a specification. Uniqueness across specifications can be accomplished using reserved dedicated namespaces or prefixes for each model, specification or domain.

Information-model specifications and standards explicitly name and define each type of the physical entity that they model. Those names, in effect, constitute a vocabulary of types with their related definitions. A *vocabulary* may be a simple list of defined names resembling a dictionary. In some specifications, the naming tree may be structured as a taxonomy to indicate possible classification relationships among the objects and to guide the proper placement of future objects as they are defined.

To facilitate unambiguous machine parsing, the vocabulary provides a precise definition of how terms are to be named and used when referring to a specific object class, such as temp or temperature. By using the type names exactly as defined in the vocabulary, communicating parties can establish an accurate correlation of references at both sending and receiving ends.

Properties and Attributes

The next component of the information model depicted in Table 6.1, object attributes and properties, are usually expressed as name-value pairs as illustrated later. Attributes generally say something about the object and its state, such as for a temperature sensor its current reading in the applicable units of measure, e.g., degrees C or F. Standards usually specify the exact format of the name fields and their meaning in the vocabulary. They also specify the data types for the values associated with the particular name fields – such as numbers, integers, or strings. Values or attributes in thing abstraction representations can be read-only, such as temperature reading, or writeable for actuation, or both. They can also include metadata, such as where the sensor is installed, what it is measuring – e.g., air or water temperature – minimum and maximum values of readings associated with the sensor, and the like.

Some standards specifications define mandatory properties that must be associated with a given type or class or objects. Some of them also define optional properties that may be associated with the specific types of objects. Since IoT servers may be implemented on constrained devices with limited processing power, some specifications and information models take that into consideration by minimizing the required mandatory set of features and actions that thing descriptions/object models must support and by supporting terse serialization variants over constrained protocols, such as binary over CoAP. A common alternative or complementary approach is to provide a proxying mechanism, where a constrained node just contains a link or a pointer to its information model whose definition is hosted elsewhere in the system.

Interactions

Some information models define the types of interactions that a modeled object supports. These can include interfaces and methods that implementations may support, such as the types of requests and responses for reading data and for issuing actuation commands. They can also include designations of ports to use for the particular protocols and formalisms, usually APIs, to interact with the thing. Device-supported APIs usually include retrieval of attributes and values, such as sensor readings and metadata. They may also include ways to request machine-readable descriptions of the thing, its object type and characteristics, or to activate the built-in methods that operate on the object's internal structures and outputs, including actuations.

Links

Finally, some sort of linking mechanism or representation is often included in IoT information models, primarily as a mechanism to point to items of interest or to form groupings of related objects. Groupings of interest may be formed to facilitate coordinated behaviors of multiple devices, such as home-automation things. They may also be used to reflect structural properties of the physical connections between devices that may be of interest to applications, such as indicating which HVAC zone a temperature sensor and an air-handling unit belong to.

Composition via linking may also be used to describe composite objects that include some more primitive basic types already defined in the specification. For example, a thermostat object type may be constructed as a composition of a temperature sensor, a set point for temperature regulation and scheduling, and an actuator of the heating and cooling system – HVAC.

Payloads and Data Serialization

As indicated, information models define the meaning and structure of device software abstraction. Endpoints involved in IoT message exchanges use those to encode and interpret the data accordingly. The messages themselves need to be serialized in a coordinated and clearly defined way, so that the receiving nodes can parse them correctly. As a generic example, the message returned by the server in response to a client's request for data may include the following:

<ObjectID>, <object type>, <attribute, value>, ..., <attribute, value>, <time stamp>

In this example, ObjectID is the name of the sourcing data instance, and object type is one from the list supported by the specification vocabulary for modeling of the related physical thing, e.g., a temperature sensor. This is followed by a list of attributes and values, such as temperature value, units of measurement, and sensor location. A time stamp is usually included to denote the time of data capture for time-series data.

Payload data serialization defines formats of requests and responses as they appear in transport. They are commonly expressed as name-value pairs also referred to as key-value pairs. JSON (JavaScript Object Notation) is one of the most commonly used forms of serialization in IoT systems [8, 9]. It uses the form "name": "value" format where *name* is the formally defined name of the property or attribute as specified in the information model, and *value* is a representation of its current state. JSON notation is humanly readable. It has a formally defined standardized syntax. As a result, software implementations of JSON validators and parsers are quite widely available and may be used to aid implementation of IoT systems. The JSON-LD (linked data) variant is popular in models that support link fields in objects [10].

In constrained IoT systems, more compact and binary representations may be preferable, such as CBOR (Concise Binary Object Representation) [11]. In legacy systems, variants of csv (comma-separated values) may also be encountered.

As a practical example of data serialization, consider an IoT server report (as a scheduled publish message or in response to a client query) that the temperature sensor had a reading of 77.6 °F at the time noted by the associated time stamp. To make the example realistic, a snippet of the server message below message is expressed more formally in the Haystack-compliant notation with JSON serialization. Haystack is one of the several standards providing IoT data models that are discussed in the chapter on standards.

```
{
"id": "150a3c6e-bef0ee0e"
"sensor": "m:",
"temp": "m:",
"curVal": "n:77.6",
"unit": "oF",
"DateTime": "t:2019-07-15T14:50:07Z UTC "
}
```

The message contains ObjectID, designated as the "id" field. In this example, it is in the form of a globally unique identifier (GUID) that the system may use to identify the particular instance of the temperature sensor. Haystack is somewhat different from other specifications in that it uses markers or tags to designate object types and metadata, denoted as "m:". In this message, two tags are used to indicate that the data is for a sensor of the type temperature, using the Haystack tag name "temp" as defined in its vocabulary [12]. The two fields functionally correspond to the object type definition in Table 6.1, for instance, defined by its "id" field. The "curVal" is the Haystack notation for the current value of the reading, 77.6 in this example. The "unit" is the Haystack name for units of measure, whose corresponding value is °F in this example. The "DateTime" field contains the time stamp associated with measurement using the format adopted by the system, in this example patterned after the ISO 8601 [13].

Being a loosely coupled distributed system, IoT endpoints should rely on the common understanding and interpretation of data payload, but they are generally not required or expected to share the same implementation. Therefore, all the required information is to be provided by the bits on the wire and whatever underlying specification for their formatting is in effect. In order to interoperate, the two parties need to comply with the common specification of object information model and its serialization on the wire such as the one illustrated above.

Metadata

Metadata or tags annotate IoT data to provide context, such as the additional attributes of a thing's functionality, placement, structural relationships with other things, and the like. Their primary function is to provide contextual semantics to create "rich data" (basically data made useful) for a variety of post-processing services and applications, such as analytics and device/asset management.

An Example of Metadata Use

Attributes and relationships described by the metadata reflect the function, states, and physical settings in which things operate. This information is required and necessary to manage control systems, such as manufacturing and large buildings. It may be implicit or explicitly stated in the system design and operational documents. An example below illustrates a rather common (and quite bad) industry practice of providing semantic annotation, in order to motivate the need for the more structured and portable metadata tagging in IoT systems that is described later in this section.

As a real-life example, consider an endpoint designation from an actual commercial building that is intended to encode its semantics:

SODA1R465--ART

This particular installation used a rather arbitrary labeling of endpoints with 14-character names, probably due to restrictions imposed by its BMS system. In this particular example, the first three letters designate a site (SOD), followed by a letter designating the type of equipment (A for air-handling unit), and followed by its number (1). This is followed by a one-letter zone designation (R), and its three-character number (465). In this case, it is followed by the two filler delimiter characters to pad the whole thing to 14 characters, ending with a designation for the ambient room temperature (ART).

Such ad hoc practices are quite common in legacy building management systems [14]. They can and often do vary from building to building, and from one system installer to another, so that even buildings that use the BMS software from the same vendor look different and require specialized training to operate. The notation is cryptic, not user friendly, and error prone. It requires system operators to consult the system "cheat sheet" of acronyms to interpret the endpoint function and location. With this approach, control applications, such as BMS, need to be customized to each particular building and its notation. This makes the resulting systems more costly, very brittle, and difficult to change to correct errors or to add improvements. Many of such systems were found to have installation errors of improperly connected or labeled endpoints that remain unnoticed for years and can adversely impact building operation and efficiency.

To indicate how the more formalized and structured metadata annotations may be used to help alleviate this problem, consider the information that is of interest to operators and applications about this endpoint. It should communicate that it:

- Is an air-temperature sensor
- Is located on the fourth floor
- Is part of an air-handling unit 1 (AHU-1)
- It belongs to an HVAC zone

Other useful information may be added, such as that the zone temperature in question is operated on a specific occupancy schedule, say occSchedule1, that is in effect from 7:30 am to 6:30 pm.

The serialization snippet below indicates how this additional metadata annotation can be expressed using Haystack-like conventions and syntax:

```
{
"id": "150a3c6e-bef0ee0e",
"dis": "Room 435 Temperature"
"sensor": "m:",
"temp": "m:",
"air": "m:",
"curVal": "n:77.60",
"unit": "oF",
"zone": "m:",
"floor": "n:4",
"scheduleRef": "occSchedule1",
"equipRef": "@AHU-1"
}
```

In this example, instance ID is designated using the Haystack name "id" with a value in the GUID-like form for machine consumption. The "dis" field is a Haystack tag, vocabulary-defined notation, for the display field that contains text for human consumption. It can be used when implementing user interfaces, such as for display on a control dashboard, and is not otherwise interpreted by the system. So, humans can think of this sensor as "Room 435 Temperature," which may be how it is designated in the installation blueprint and on the dashboard, and machines can use the GUID id field for unambiguous identification. The next few metadata fields are similar to what we have encountered before – in this case air temperature sensor, reading 77.6 °F, located in a (HVAC) zone.

The last two entries are references to other items whose function is similar to the "links" field introduced in the IoT object structure in Table 6.1. In this example, "scheduleRef" points to "occSchedule1," which is a document that can be located through occSchedule1machine-readable definition document not shown here. AHU-1 and occschedule1 definitions are not shown in this example, but they are assumed to be provided elsewhere in the system description. The "equipRef" is Haystack reference mechanism for expressing structural links, such as that the temperature sensor in question is in the HVAC zone controlled by the AHU-1.

This information may be used by operators and applications to determine whether the reading is in the desired range, by comparing it to what is specified in schedule 1 for the time of the day when the reading is taken. If it is in range, no action is needed. If not, the related air-handling unit, AHU 1 in this example, can be actuated to cool or to heat the related zone, as appropriate, until the target setting is reached.

The key takeaway here is that the structured metadata annotation and use of controlled vocabularies can enable applications to infer attributes of interest by querying. The obtained machine-readable and understandable descriptions are structured in a commonly documented format that is system-independent, i.e., should work for all systems and compliant installations in the domain covered by the specification.

This facilitates the creation of portable IoT applications that can work on many systems, which increases their value and lowers implementation costs.

Types of IoT Metadata

Metadata can come in a variety of forms. As an illustration of metadata diversity and range, below is a partial list of IoT sensor data and metadata grouped into some representative categories with *italicized* names. This example omits endpoint reading (values) and writing (actuation), because they are technically data which metadata can annotate.

Sensing point

- Units of measurement
- Range – min, max values
- Resolution, sampling rate
- Frequency of reporting
- Scaling factor
- Time stamp

Location

- Location – static or dynamic if sensor is mobile

Maintenance and asset management

- Status = {active, inactive, connected, fault}
- Battery status
- Accuracy, calibration
- Manufacturer, part#, serial#
- Configuration settings
- Firmware, software version installed

Access rights, privacy, security
Owner/domain and associations
Reputation
Other, extensions, etc.

Location is a very important aspect of metadata. For static sensors, such as thermostats, it may be useful to know the geographic location, such as the GPS coordinates or street address, as well as indoor location, such as the floor and the room where it operates. This type of metadata is static. It rarely if ever changes during the life cycle of the thing, so it need not be reported with every reading. For mobile sensors, such as those in vehicles and smart phones, annotating each reading with the GPS location of its capture is often valuable.

Some metadata fields that describe intrinsic attributes of a thing – such as the range for temperature sensor readings, accuracy, manufacturer, and model – may be defined at the manufacturing time and included as a part of its instance definition depicted in Table 6.1 in the attributes/fields area. Other metadata annotations – such as its location within a building – can be defined later in the system life cycle, at system installation and commissioning time. An IoT system needs to provide a mechanism for adding data annotations, metadata, at various times in system life cycle; for associating metadata with the thing that it relates to, using its identifier; and for retrieving data and metadata in response to service requests.

In general, much of the metadata tends to be relatively static or to change much less frequently than the associated data. This, fortunately, can be used to reduce the size of the endpoint reports as only the metadata that has changed since the last reported value needs to be included. In practice, this can be accomplished by establishing a shared contextual reference of metadata entries and values between the edge (gateway or smart sensor) and the cloud. Metadata values can be initially provided at the system configuration and provisioning time and communicated only when changes occur. It is also useful to provide a mechanism and an API to request an endpoint's data in its entirety. This can be useful for system checks in situations such as recovering from the catastrophic node failures or when connecting to a different back-end system or cloud. An example of how this can work in practice is provided in chapter "Putting It All Together".

In addition to the contextual annotation, an important aspect of using structured metadata in IoT systems is to enable attributed-based searches. In general, the primary purpose of a search is to identify objects of interest based on the properties specified in the search request. The requesting entity does not know a priori which objects those are. It issues a search query that returns qualifying endpoint IDs for the subsequent retrieval.

IoT data are essentially numbers and states that cannot by themselves be indexed in interesting ways. For example, it is not very useful to find all or a subset of IoT data whose value is "close to 30.5." To be searched in meaningful ways, IoT data need the context and semantics that are provided by the metadata.

When metadata are provided, IoT systems can support attribute-based searches of its data, such as by type, owner, location, reading value(s), and time period. For example, a building-management compliance application may request "all available temperature readings on the fourth floor at 2 pm." With the proper metadata annotation, such sensors will be tagged with floor and perhaps room locations and support such searches. Attribute-based searches are immensely useful to applications and services, and, when supported, they can significantly increase the value of an IoT system. On the flip side, IoT systems without metadata allow only extremely limited forms of data retrieval, such as the reading of individual objects identified by their name and possibly a specific time.

IoT Frameworks

IoT information models as described thus far are necessary but not sufficient for achieving full operational interoperability in IoT systems. Namely, in addition to using the same information model, in order to be able to interoperate, IoT endpoints also need to use the common set of protocols and serialization mechanisms, compatible naming and addressing conventions, and operate in the common security perimeter. IoT frameworks are intended to provide full operational interoperability by extending their scope of definition beyond the information model to include specification of the complete operating environment for the compliant IoT nodes. In addition, frameworks may include bridges and protocol converters to facilitate interoperability with the devices and things that are using conventions and legacy protocols that may not be Internet compatible. As indicated in chapter "Communications", IoT frameworks generally operate above the transport layer but below the application layer.

IoT and M2M frameworks may define some or all of the following:

- Data (information) model
- Payload serialization
- Protocol bindings
- Naming and addressing, discovery
- Security
- Device management and provisioning

Framework definition of data and information models typically include object type definitions, properties, and interactions, with types of requests and responses, supported interfaces, methods, and access points for those, such as the specific port numbers.

Protocol bindings specify which protocols are supported and perhaps the port numbers on which they operate. In addition to the common Internet protocols, more compact versions, such as CoAP, may be favored in the constrained segments of the network.

IoT frameworks usually define the conventions and mechanisms to be used for the thing naming and addressing. Object name is often correlated to its addressing mechanism, i.e., knowing the name typically can be used to locate the object. Depending on the IoT specification and capabilities of the thing, object name may be an Internet-style Universal Resource Identifier (URI), a Universal Resource Locator (URL), or a Universal Record Name (URN), specifying its name, location, or both. The minimum requirement for an ObjectID is to be unique within a domain in which it resides and correlated to the thing naming and addressing mechanism that may be implemented elsewhere in the domain, such as in a separate thing directory.

The Internet practice and convention is to use the resource endpoint names and identifiers that are resolvable to IP addresses by its domain naming system (DNS) that acts as a form of distributed directory. Due to the complexity, cost, and limitations of this system (which was not designed with billions of IoT endpoints in mind), some IoT frameworks opt to include their substitute or supplemental mechanisms that work only within the specific domains. This may be accomplished with the use of internal intra-domain device and thing directories and registries. They can also be used for intra-domain node discovery, which can be more efficient and secure than exposing them directly to the Internet. Global reachability of the domain nodes of interest may be achieved by using the reserved domain names as a prefix to direct the traffic to the relevant domain, where it can then be additionally resolved to an internal address.

Object name is often correlated to its addressing mechanism, i.e., knowing the name typically can be used to locate the object. Depending on the IoT specification and capabilities of the thing, object name may be an Internet-style Universal Resource Identifier (URI), Universal Resource Locator (URL), or Universal Record Name (URN), specifying its name, location, or both. The minimum requirement for ObjectID is to be unique within a domain in which it resides and correlated to the thing naming and addressing mechanism that may be implemented elsewhere in the domain, such as in a separate thing directory.

Frameworks may also specify security requirements for compliance and node connectivity. These may include the required and sanctioned types of authentication, authorization, encryption, and security protocols, such as TLS.

Some frameworks also specify elements of device management, such as the forms of health and status reporting, software and security updating, and methods for initial device on-boarding and provisioning.

As indicated, IoT frameworks can define many aspects of an IoT system, well beyond the information model. Nodes wishing to interoperate within a framework must implement all of its applicable components. The positive side of using frameworks is that they can assure interoperability at virtually all levels of an IoT system. As a consequence, some framework implementations also offer formal certification of product compliance that can be tested by the authorized entities. In addition to the specification, some frameworks are implemented as middleware with tested and sanctioned implementation that can work on multiple platforms and can aid system implementations.

On the negative side, use of a framework requires adherence to all of its mandatory parts which may be a problem when its specification is at odds with the design preferences and requirements of a particular system. Another practical problem is that framework compliance limits system implementations to only those thing types that are defined in its specification. In current specifications, those are limited to not much more than 100 object types in particular domains, which can be a serious limitation given the much greater and ever-growing number of thing types in IoT systems.

IoT Interoperability

Previous sections introduced the notion and use of IoT information models to achieve interoperability. This level of interoperability works only among the compliant components that use and implement a common shared specification. This is in effect an intra-specification or intra-domain interoperability. It does not work across nodes that may reside in different domains and use a different specification [15]. In this and the subsequent sections, we focus on semantic IoT interoperability

across domains and specifications and discuss some of its characteristics and implementation directions. To do this, we refine the definition of interoperability and
introduce some insights and concepts that frame the problem in a way that can be
pragmatically addressed in practice.

When supported, semantic interoperability across domains achieves several
important objectives:

- Allows large aggregations of (useful) data from different domains
- Enables IoT system integration across components and vendors
- Enables integration with legacy data
- Supports creation of portable IoT applications

A significant benefit of IoT data interoperability is the ability to create large, useful aggregations of sensor data for post-processing such as data mining, analytics,
machine learning, and AI. It is well known that effectiveness of all those techniques
increases with the size and diversity of data sets. At present, IoT data are fragmented
and locked in silos due to incompatibility of formats in proprietary platforms and in
numerous evolving standards that focus on limited domains. A great opportunity
lies in being able to create horizontal services and applications that make use of
aggregations of sensor data across devices and domains. Such services can focus
more on the needs and interest of users, e.g., as data from devices belonging to a
user or of interest to the user in a particular context, surrounding, or at a given point
in time.

Data interoperability is required to enable creation of systems integrated by
mixing and matching components from different vendors, such as web servers and
browsers that is the norm with the Internet. Due to the lack of a common data-
model specification and interoperability, in today's practice it is generally not possible to integrate IoT end-to-end systems constructed using components from
different vendors, such as a variety of gateways running different middleware, with
a third-party cloud for data storage, processed by the analytics from independent
vendors.

Another significant benefit of an architected approach to interoperability is the
ability to incorporate volumes of existing sensor data being produced by the legacy
systems, including proprietary Supervisory Control and Data Acquisition (SCADA)
systems, building management systems (BMS), energy utilities, and various automation and manufacturing systems using legacy standards such as BACnet
and Modbus.

In addition to increasing the IoT data set size and diversity, the availability of a
commonly understood format enables the creation and use of portable IoT applications and services – such as visualization, analytics, and AI – and significantly
accelerates variety and usefulness of offerings by following the proven Internet
playbook. In the absence of service-level data interoperability, IoT services and
applications tend to be custom tailored to specific systems and data formats, which
makes it difficult and, costly, and time-consuming to port them to another system or
cloud hosting service.

Levels of IoT Interoperability

Computing and engineering literature tends to define three levels of interoperability: (1) syntactical interoperability, (2) structural interoperability, and (3) semantic interoperability [16]. The basic notion is that with syntactical interoperability, endpoints can parse the data fields; when structural interoperability is achieved, they can understand the structure of data and with semantic interoperability understand the meaning of data. While keeping these in mind, this section focuses on the pragmatic aspects of defining and achieving semantic interoperability in settings commonly encountered in IoT systems.

The current state of affairs in IoT information modeling is that there are many different standard proposals and proprietary specifications with incompatible data models and definitions, usually aimed at rather narrowly focused application domains. Several of the proposed standards are described in chapter "IoT Data Standards and Industry Specifications". Given a relatively large number of IoT domains and a vast number of things therein, it is unlikely that any single unifying IoT information or even data model standard will emerge in the foreseeable future. In the meantime, some pragmatic and rational simplifications can be used to manage the problem and to achieve some significant IoT interoperability benefits.

There are three major and somewhat distinct levels and flavors of interoperability in IoT systems that have different aims, requirements, and practical solutions. They are:

• Intra-domain interoperability, machine-to-machine, usually at the edge
• Inter-domain interoperability, typically encountered at intermediate levels of IoT system hierarchy
• Multi-domain interoperability for processing large aggregations of diverse data, typically in the cloud

These three basic levels are depicted in Fig. 6.2. The figure is purposefully skeletal to illustrate key points; multiple levels of each layer can be instantiated in complex IoT systems.

Figure 6.2 depicts several edge devices operating in two different domains, each with its aggregation and control point, labeled as ACP. An ACP may aggregate, filter, and store data and operate group-control algorithms and services, such as analytics, that make use of data and state provided by the edge nodes in its domain. Such settings are common in complex hierarchical control systems that may implement several intermediate levels that are not shown in Fig. 6.2 for simplicity. Note that the ACP nodes are conceptual and can be optional in the sense that a number of IoT edge devices may operate in a domain in a purely peer-to-peer fashion, without a centralized node or a controller. In such configurations, edge devices coordinate their behaviors to achieve some mutually agreed common purpose for which the domain is formed, e.g., adaptively manage ambient conditions in a home environment. In so doing, they are effectively implementing a distributed version of data aggregation and control.

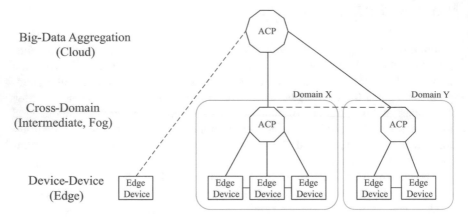

Fig. 6.2 Levels of IoT data interoperability

Before discussing IoT interoperability in such settings, we cover the special case of the edge-to-cloud mode of operation which is somewhat of an outlier but is quite common in certain applications. It is depicted in Fig. 6.2 by a single edge device that is functionally connected to a cloud service.

This mode of operation is typical for some consumer devices, such as the personal activity monitors. They may contain a battery-powered wearable that collects data and does some processing and buffering before communicating data to the matching cloud service, directly or via an intermediary, such as the smart phone. The smart phone typically runs an application that displays user data and allows setting of preferences and communicates data to a cloud service where it can be processed and archived. Aggregated user data can be subjected to multi-user analyses, such as comparison of activity with friends and competitors, and computing various kinds of statistics of interest to users and the service provider.

In practice, these are often closed systems that require users to run a vendor-provided phone application and to rely on their cloud service for post-processing. In such settings, data formats are usually proprietary. If at all, they can be accessed externally only through the vendor-controlled cloud APIs. Typically, only limited access is possible in such systems without the special arrangements or licensing from the vendor. For users, such closed systems and the resulting lack of interoperability are inconvenient for several reasons. For one, it forces them to rely on the dedicated apps for almost every one of their IoT devices. This is sometimes referred to as the "app trap." Another problem is that the collected data tend to be owned by the service provider and not the user. Thus, if a user decides to start using a new model of a wearable device from a different vendor, months or years of the personal activity logged by the prior vendor will not be combinable to provide a cumulative continuity that the user would likely want to have. From the Internet point of view, the loss is that the individual user data are locked in vendor silos and are not available for the big-data aggregations that could be used to discover population mobility patterns and trends and perhaps be used for the common good such as health studies.

Intra-domain Interoperability

As illustrated in Fig. 6.2, groups of devices often operate within a specific domain, such as a home, a factory production line, or a building management system (BMS). Domains are formed among collections of devices to achieve some common objective, such as production quality, or energy efficiency and occupant comfort in a large commercial building. They are typically tightly coupled systems operating within a common security perimeter and with a shared information model defining syntax and semantics for data and control exchanges. Most of the current IoT standardization efforts and proprietary legacy control systems fall into this category, albeit with different approaches – open vs. closed – to data definition and metadata handling.

Devices that are part of a domain usually achieve full interoperability by sharing a common information-model specification and, where applicable, the same framework. Full intra-domain interoperability is achieved through the compliant implementations of the shared specification by all devices in the domain. This implies that devices in the domain typically can discover the others and exchange data and control messages in a peer-to-peer fashion. Both syntactic and semantic interoperability are achieved through a common specification. It defines the types of messages that can be exchanged, how they are formed (syntax), and what object types that represent real-world things are and therefore mean (semantics).

Inter-domain Interoperability

Implementation of inter-domain interoperability can be simplified through the realization that not every device in one domain needs to be able to exchange data and controls with any device in another domain. This would be not feasible without sharing the common information model. Fortunately, it turns out that there are other simplified forms of interoperability that are very useful in practice and that do not require existence of the common information model.

For example, in large commercial buildings security, fire alarm systems, BMS, and elevator controls tend to be provided by different vendors. They usually have very different and proprietary forms of data representation and information models, thus creating rather opaque data and control silos.

With some level of interoperability, their coordinated behaviors can be beneficial for improving the occupant comfort and safety, as well as the building operation and energy efficiency. For example, security systems that track the entrance and exit of individuals through gate automation can and often do calculate accurate building occupancy. Since people are the primary load and drivers of ambient-condition requirements in large office buildings, timely occupancy data can be used by the BMS to adapt use of HVAC systems to the actual and dynamically changing building needs – as opposed to the common practice of schedule-based management that is neither adaptive nor efficient. More importantly, in case of emergency such as triggering of a fire alarm, cross-system communication can lead to desirable coordinated system behaviors, such as the elevator cars being moved to the base floor, BMS set to the fire conditions in the affected areas, exit gates open, and occupancy tracked to know if and how many people are still left in the building.

As these simple examples indicate, inter-domain interoperability is useful even when applied only to the aggregate, group-level data. For example, it is useful for the BMS to be able to query or receive notifications from the security system when certain occupancy thresholds are reached. However, it is not necessary for every device in the BMS domain to talk to every device in the security system. For example, a thermostat in room 3 on the fourth floor has no need to talk to the entrance count on the badge-operated gate number 5. Nor does it need to talk to a flow sensor #117 on the Alaskan Prudhoe Bay pipeline or to a location sensor on the cow #5732 in the Netherlands.

In other words, IoT systems and design should strive for the functional interoperability of data and events that are relevant. With the current state of affairs, it is neither practical nor necessary to aim for the complete interoperability in the sense of enabling any IoT device/thing in a domain to talk to each and every other IoT device/thing in another domain directly. This realization is a key insight in motivating what might be called the rational interoperability – support only those cross-system interactions that are necessary and useful, not anything and everything that could possibly be done. With numerous proprietary and domain-specific IoT standards in existence and under development, the challenge to IoT community is to work on a more rational approach to interoperability [17].

Multi-domain Interoperability for Big-Data Aggregation

Large aggregations of IoT data, shown at the top level in Fig. 6.2, serve a different purpose. Their primary aim is to provide large pools of diverse data for the more general system-wide or global correlations and pattern and trend analyses through the use of analytics and AI. For example, data from wearable sensors from a large and diverse population of users can indicate general population activity and perhaps yield additional insights when cross-referenced with the medical data. An extensive collection of energy efficiency data on buildings can be used to create practicable guidelines and operational targets by analyzing behaviors of buildings of similar size and construction in similar climates. Such aggregations are not possible today due to the lack of data interoperability.

Large data aggregations, cited as a major benefit of IoT systems, are most useful when they are diverse. This implies that their data are sourced from multiple domains. Technically this is also inter-domain interoperability but intended for the large, multi-domain and even global aggregations of data. In such setting, the interoperability challenge can be simplified by the fact that few if any cross-domain system interactions may need to be supported. Data at rest with their semantic annotations are of primary interest in large aggregations. The ability to interact directly with a data source is typically not required, as some intermediary agent has already done that to acquire the data and bring it to the pool. Therefore, interoperability definitions and translations may be considerably simplified by omitting the interaction portion of the information models that are being aligned. This can reduce the problem to primarily establishing equivalences between the vocabularies of the target models.

Rational interoperability at this level calls for cross-application and cross-domain understanding of object types and relevant metadata semantic annotations. In practice, it is somewhat simplified by the fact that most IoT data are stored as name-value pairs in JSON-like formats. Moreover, information models represent things whose semantics are objectively the same as defined by their physical characteristics. Therefore, models that reflect them should be structurally similar and the problem is largely reduced to establishing the equivalence of vocabularies of object types and attributes across different models. Several modeling efforts are working on meta-model representations that would capture the salient features of modeled things and enable machine translations between the specific representations. Some of them are described in chapter "IoT Data Standards and Industry Specifications".

In the meantime, some more pragmatic work on the curated inclusion of semantic annotation is being done in schema.org [18] and more topically for IoT in iotschema.org [19]. Taxonomies and ontologies may help in this process, but their practical implementation at the scale required to describe potentially all of IoT things is a formidable challenge.

IoT Practitioner's Data Modeling Navigation Guide

As indicated in chapter "IoT Data Standards and Industry Specifications", IoT standards and published information models at present are relatively narrow in scope, limited to specific domains, and very much in development. Practitioners who are designing or integrating production IoT systems in the interim may have to be prepared to work with several information models [20] and possibly design their own. When adding IoT to existing systems, they will likely also encounter legacy components with proprietary data formats that may need to be brought into the IoT fold.

For smaller pilot projects and proof-of-concept explorations of various aspects of IoT whose data may be discarded upon planned completion, elaborate data formats and information modeling may not be of primary concern. However, if the collected data are intended to be used later, a more considered design should be undertaken.

When IoT is applied to production systems, it is often desirable to be able to retain the collected data across the likely changes, such as system extensions or migration to a different implementation. At the very least, archived data can provide a useful historical perspective that is beneficial for the operation of prediction algorithms and analytics.

An IoT data information model should contain a consistent and documented vocabulary for naming and definition of the supported thing types and attributes, including metadata.

When relying upon or extending an existing standard, system designers should follow its architectural style and assumptions when defining the types of things that are not included in the specification but need to be supported by the system being designed. When the custom definition route is chosen, it should follow the general information model design principles and practices described in conjunction with Table 6.1. In either approach, device interactions and access points should be formalized and documented, as appropriate.

While not part of the information-model definition, practical interoperability requires explicit articulation of the environmental assumptions for the operation of things and services in an IoT system, including connectivity framework, communications protocols, naming and addressing, security, discovery, etc.

Planning for the future and a broader use of collected data should include accommodations for inter-domain interoperability and fitness for big-data aggregations using articulated guidelines, such as the rational considerations described earlier.

References

1. McKinsey Global Institute Report (2015) 'Unlocking the potential of the internet of things' [Online] Available at: https://www.mckinsey.com/business-functions/mckinsey-digital/our-insights/the-internet-of-things-the-value-of-digitizing-the-physical-world (Accessed Dec 15, 2019)
2. IIC The Industrial Internet of Things Volume G1: Reference Architecture (2019) [Online] Available at: https://www.iiconsortium.org/pdf/IIRA-v1.9.pdf (Accessed Dec 15, 2019)
3. Semantic Web [Online] Available at: https://www.w3.org/standards/semanticweb/ (Accessed Dec 15, 2019)
4. Semantic sensor network ontology [Online] Available at: https://www.w3.org/TR/vocab-ssn/ (Accessed Dec 15, 2019)
5. W3C Resource Description Framework (RDF) [Online] Available at: https://www.w3.org/RDF/ (Accessed Dec 15, 2019)
6. W3C Web Ontology Language (OWL) [Online] Available at: https://www.w3.org/OWL/ (Accessed Dec 15, 2019)
7. Hodges, J., Garcia K and R. Steven (2017) 'Semantic development and integration of standards for adoption and interoperability' *IEEE Computer* 50(11), p26–36.
8. Bray, T. (2017) 'The Javascript object notation (JSON) data interchange format', *IETF RFC 8259* [Online] Available at: https://tools.ietf.org/pdf/rfc8259.pdf (Accessed Dec 15, 2019)
9. ECMA standard 404: the JSON data interchange format [Online] Available at: http://www.ecma-international.org/publications/files/ECMA-ST/ECMA-404.pdf (Accessed Dec 15, 2019)
10. JSON linked data specification, JSON-LD [Online] Available at: https://json-ld.org/spec/latest/ (Accessed Dec 15, 2019)
11. Bormann, C. and P. Hoffman (2013) 'Concise binary object representation' IETF RFC 7049 [Online] Available at: https://tools.ietf.org/html/rfc7049 (Accessed Dec 15, 2019)
12. Haystack Tags [Online] Available at: https://project-haystack.org/tag (Accessed Dec 15, 2019)
13. ISO 8601 Date and Time Format [Online] Available at: https://www.iso.org/iso-8601-date-and-time-format.html (Accessed Dec 15, 2019)
14. Bhattacharya, A. et al. (2015) 'Automated metadata construction to support portable applications in buildings' *Proceedings of the 2nd ACM International Conference on Embedded Systems for Energy-Efficient Built Environments' (Buildsys),* Seoul, South Korea, p 3–12.

15. Milenkovic, M. (2015) 'A case for interoperable IoT sensor data and meta-data formats' *The internet of things', (Ubiquity symposium), ACM Ubiquity*, Nov.[Online] Available at: https://ubiquity.acm.org/article.cfm?id=2822643 (Accessed Dec 15, 2019)
16. Wang, W., A. Tolk and W. Wang (2009) 'The levels of conceptual interoperability model: applying systems engineering principles to M&S' *Proceedings SpringSim '09* [Online] Available at: https://arxiv.org/ftp/arxiv/papers/0908/0908.0191.pdf (Accessed Dec 15, 2019)
17. Schooler, E. M. et al (2018) *'Rational interoperability: a pragmatic path toward a data-centric IoT'* 2018 IEEE 38th International Conference on Distributed Computing Systems (ICDCS), Vienna, pp. 1139-1149.
18. Schema.org [Online] Available at: https://schema.org/ (Accessed Dec 15, 2019)
19. Iotschema.org [Online] Available at: http://iotschema.org/ (Accessed Dec 15, 2019)
20. IETF Thing-to-Thing Research Group [Online] Available at: https://datatracker.ietf.org/rg/t2trg/about/ (Accessed Dec 15, 2019)

Chapter 7
IoT Data Standards and Industry Specifications

There are many standards that are applicable to various parts of IoT systems. Since IoT things usually operate in M2M environments and are also part of the Internet, many of the standards designed for those environments may be directly or indirectly relevant to IoT systems. In this chapter we briefly describe several representative and influential ones that focus on data and information models that are of particular importance to IoT. Some of them are created by industry groups and associations that are not formal standards bodies and tend to be more generically referred to as the Standards Definition Organizations (SDOs).

There are a number of standards and industry specifications that provide IoT information models and frameworks to a different extent. They have similarities and differences, many of which are illustrated by the representatives described in this chapter. They all model physical objects, such as sensors and actuators, into software representation that servers who are hosting instances of the resources can use to reflect and report their states. Consuming applications and services, IoT clients use the data models to create their own instances of thing representations and to interpret responses from servers. Use of the common specification results in syntactic and semantic interoperability in the sense of interpreting what the data means.

The five standards covered in this chapter are IPSO (IP for Smart Objects), Haystack, OCF (Open Computing Foundation), WoT (Web of Things), and OPC (Open Platform Communications). IPSO is included as one of the first examples of a pure object-based IoT data model. Haystack illustrates a comprehensive and flexible scheme for incorporating metadata descriptions through crowdsourcing. OCF prescribes an entire IoT framework, including an information model with actions and links as well as protocol bindings, security, and management. OCF also provides machine-readable specifications, tools for crowdsourcing, a reference implementation, and compliance certifications for products that implement it. Web of Things is the work of the World Wide Web Consortium (W3C). It does not provide a definition of an IoT model, but a formal specification on how to describe other models in a manner that facilitates interoperability, aiming to be a conceptual equivalent of HTML for things. OPC is an example of an older legacy specification for

© Springer Nature Switzerland AG 2020
M. Milenkovic, *Internet of Things: Concepts and System Design*,
https://doi.org/10.1007/978-3-030-41346-0_7

industrial automation that has the industry following and an installed base of users and compliant products. As such, it continues to be a factor in creating interoperability across domains and in data aggregations.

The chapter closes with observations on similarities and differences and on some issues and limitations of IoT data standards.

Terminology in IoT Data Specifications

Specifications included in this section are representative and, in many ways, illustrative of a larger number of standards. One of the unfortunate commonalities is that IoT standards and specifications can be difficult to read and compare with each other. Some major problems include the lack of explicitly articulating their scope, intent, and environmental assumptions. For example, specifications can range from pure data and information models to complete frameworks. In either case, it is useful to understand their environmental assumptions, i.e., what in addition to what they specify is assumed to be in place for the adopting IoT system to be implemented in practice – such as communications stacks, protocols, naming, and security. Another important thing to understand is the target domain of the specification and object types that it defines. For example, it could be consumer, manufacturing, or smart buildings. This is rarely made clear, as if the specifications were produced in the belief that they should be universal. In practice, the reader may be forced to deduce the target domain by inspecting the nature of object types that are actually defined in the specification.

However, terminology is probably the worst offender. Different specifications use different terms for common concepts, such as object, resource, and property. As indicated in Table 7.1, the terminology used for model representations varies among the specifications described in this chapter. While the result is fairly simple, it was derived from a lot of readings and comparisons of different specifications and from discussions in various standards groups and meetings. Assigning different meanings to the same words when referring to the same set of concepts is a recipe for confusion and frustration. This makes for sometimes diverging conversations in meetings involving representative of different groups, as they often talk past each other by using the same words to describe different concepts. As the saying goes, the most common mistake regarding communication is the assumption that it has occurred.

Table 7.1 IoT standards terminology

Generic	IPSO	OCF	Haystack	Web of Things (WoT)	OPC UA
Physical thing (modeled)	Object	Resource	Entity	Thing (description)	Object
Attribute, value	Resource	Property	Tag	Property	Property
Event	n/a	Event	n/a	Event	Event
Others				Action	

For consistency, in general discussions we try to refer to physical entities as "things" and to their modeled representations as "thing abstraction." Individual specification sectional descriptions will use the native terminology of the standard being described, with Table 7.1 providing a decoder ring for comparative interpretations. This is intended to facilitate correlation with the terminology used in following the individual specification documents.

IPSO (IP for Smart Objects)

The IPSO (IP for Smart Objects) alliance was formed by a group of industrial members in 2008 to enable "smart object communication," essentially machine-to-machine (M2M). One of the founding tenets of IPSO was that things should have IP endpoint addressability, which was a matter of considerable debate among the early implementers of constrained IoT devices.

IPSO provided one of the first definitions of data objects and elements of information model prototypical for IoT standards. It can be viewed as somewhat of a pure data model in the sense that it primarily defines object classes and their properties, called resources in IPSO documentation.

The IPSO specification is focused on data objects and intended to work at framework and application layers. It does not specifically cover security, management, and provisioning. For those, IPSO is meant to provide bindings to other specifications. The only published IPSO binding is the Object Management Association (OMA) Lightweight M2M (LWM2M) specification that also includes management, security, and designation of lower-level protocols, such as CoAP. The two specifications are tightly coupled in the sense that IPSO objects are registered in LWM2M OMA Naming Authority (OMNA) and LWM2M refers to IPSO for its data and object model specification. The two organizations have merged in 2018 to provide a more complete IoT framework and definition stack, OMA SpecWorks [1, 2].

In IPSO terminology, an object representation structure includes an object-type definition, such as temperature, and a collection of formally named resources, each of which represents a simple value that exposes or describes a function of the object [3].

IPSO specifications provide the object model and definitions for smart things that follow a structured design pattern. Their use is intended to provide high-level interoperability in exchanges between smart IoT devices and connected software applications on other devices and services. The IPSO object model has defined OMA LWM2M bindings and can optionally be used with other protocols, such as HTTP, that support the web standard content types and REST methods.

Objects and resources are mapped into the URI path hierarchy in which each URI path component sequentially represents the object type ID, the object instance ID, and the resource type ID. The structure consists of three unsigned 16-bit integers separated by the character "/" in the following form:

Object ID/Instance ID/Resource ID

This URI template approach follows the Web Linking [4] and the IETF CoRE Link Format [5]. Objects are typed containers, which define the semantic type of instances. Object instances are containers for resources, which are the observable properties of an object.

An example of URI for a temperature sensor in IPSO is

3303/0/5700.

"3303" is an IPSO object ID for the temperature sensor, "0" designates the instance ID, and "5700" is the resource ID for the sensor value, called property in IPSO terminology.

Semantically, the object type represents a single measurement, actuation, or control point. The resource specifies a particular view or active property of an object. For example, a temperature sensor object might expose the current value, the minimum and maximum possible readings, the minimum and maximum reading in an interval, and the attributes like engineering units. Semantic interoperability is achieved through a pre-defined set of object and resource identifiers.

Naming of objects and properties uses URN (Uniform Resource Name) that is unique to IPSO/OMA, as registered and specified in OMA namespace. URNs have the format "urn:oma:lwm2m:{oma,ext.,x}:{ObjectID}," where the part in curly braces {} means that those values are variable and filled with real values. For example, an object URN for a temperature sensor is "urn:oma:lwm2m:ext.:3303" where 3303 is the object ID for (IPSO) temperature object specified by the OMA.

Therefore, IPSO would represent the temperature reading of temperature sensor instance 0 (other temperature sensors, if present, would be instances 1, 2,...) of 42.34 as

3303/0/5700 42.3,

and its minimum and maximum allowable values would be designated, respectively, as

3303/0/5601 34
3303/0/5602 45

followed by numbers representing values of those, such as 34 and 45.

IPSO also defines reusable object properties that may be applied to compose objects. Those properties are called reusable because they are fairly common in smart objects and may be applied to multiple object types. These properties are defined in the Reusable Resource Registry that is maintained by OMNA. There are over 50 of them covering a variety of properties, such as the state and polarity of digital and analog inputs, counters, minimum and maximum measured values and range, active and reactive power values and ranges, duty cycles, and set point values. Six of them are shown in Table 7.2 as associated with the temperature object.

IPSO has defined over 50 types of smart objects. They include digital and analog inputs and outputs, generic I/O sensors, presence, temperature, humidity, power, load control, light control, barometer, accelerometer, actuation, and set point.

Table 7.2 IPSO temperature object definition [6]

Temperature

Description

Description: This IPSO object should be used with a temperature sensor to report a temperature measurement. It also provides resources for minimum/maximum measured values and the minimum/maximum range that can be measured by the temperature sensor. An example measurement unit is degrees Celsius (ucum:Cel).

Object definition

Name	Object ID	Object Version	LWM2M Version
Temperature	3303	1.0	1.0
Object URN		Instances	Mandatory
urn:oma:lwm2m:ext:3303		Multiple	Optional

Resource Definitions

ID	Name	Operations	Instances	Mandatory	Type	Range or Enumeration	Units	Description
5700	Sensor Value	R	Single	Mandatory	Float			Last or Current Measured Value from the Sensor
5601	Min Measured Value	R	Single	Optional	Float			The minimum value measured by the sensor since power ON or reset
5602	Max Measured Value	R	Single	Optional	Float			The maximum value measured by the sensor since power ON or reset
5603	Min Range Value	R	Single	Optional	Float			The minimum value that can be measured by the sensor
5604	Max Range Value	R	Single	Optional	Float			The maximum value that can be measured by the sensor
5701	Sensor Units	R	Single	Optional	String			Measurement Units Definition e.g. "Cel" for Temperature in Celsius.
5605	Reset Min and Max Measured Values	E	Single	Optional				Reset the Min and Max Measured Values to Current Value

IPSO objects and their resources have the same operations as their counterparts in the OMA LWM2M specifications. For example, objects and their instances can be read and written. Resource values can be read, written, or executed for some restricted types. Content formats for serialization are also specified, and they include text, JSON with SenML, CBOR (Concise Binary Object Representation), and binary.

Temperature Sensor in IPSO

In order to illustrate the style and general appearance of various specifications discussed in this chapter, we use as an example a common and fairly straightforward temperature sensor. We begin with an excerpt from its definition in the respective specification, followed by a snippet of a serialization that could represent a server's response to a request for its value report. Representations are based on some published version of each specification. Since specifications are evolving and changing, the provided examples should be used for illustrative purposes only.

A typical specification for a temperature sensor includes a class name for the object type, say "temp," an instance ID, a property for reporting the current reading, possibly its data type such as floating-point number, and optional metadata such as units of measure. Many specifications also include variants of properties related to range and minimum and maximum values. These may be defined by the designed specification of the range that the physical sensor is rated for, min and max values of the normal operating range, or over a period of time until they are reset. They are intended to enable applications to detect deviations and to generate appropriate events.

In this example, it is assumed to be the temperature in a chilled water pipe that is supposed to remain within the stated range of (34, 45) ^{0}F. Those values would be specified by the designer of the cooling system. The receiving application could use them to check that the temperature measured by the sensor is within the specified range, as is the case in this example, or detect a malfunction and generate a corresponding event if it is not.

Table 7.2 is an IPSO definition of the temperature object. It starts with an Object ID (3303), followed by a list of reusable properties that may be applied to it. The only mandatory property is the sensor value (5700). Optional properties include minimum and maximum measured values, range, and sensor units of measurement.

The code in Fig. 7.1 is an IPSO example of a temperature report.

The serialization is not explicitly defined in the original IPSO specification of smart objects. In Fig. 7.1 we use JSON-like notation with "n" indicating names and "v" values. The IPSO temperature object is designated as the prefix to all other values, which is "3303/0," for instance, "0" of the temperature sensors (IPSO instance count starts with 0). Other fields represent the mandatory value property (resource in IPSO as specified in Table 7.2), unit of measurement (resource 5701), and the optional minimum and maximum values designated as reusable properties 5603 and 5604, respectively.

Haystack

Haystack, also referred to as Project Haystack, is an industry association of companies that provide system equipment and applications and services for building management systems, such as automated control, visualization, and analytics [7, 8].

```
"prefix": "3303/0/",
    {
              "n": "5700",
              "v": 42.3,
              "n": "5701",
              "v": "F",
              "n": "5603",
              "v": 34
              "n": "5604",
              "v": 45
    }
```

Fig. 7.1 Temperature sensor report in IPSO representation

While its origins are in the domain of buildings, the Haystack philosophy and architectural approach are applicable to IoT systems in general. Haystack's primary technical contribution of interest here is the versatile, extensive, pragmatic handling and annotation of metadata, called tags.

Haystack provides a common methodology for defining metadata through a set of named tags to express attributes and properties that may be associated with system entities. Tags are generally expressed as name-value pairs that describe facts about those entities. Haystack defines a common vocabulary in the form of libraries of community-defined tags. Tags have fixed names and the libraries define their associated meanings. Applications and services that consume the data can interpret the meanings of tags by using the same Haystack libraries that data are annotated from.

One of the interesting aspects of the Haystack approach is that metadata annotations, i.e., tags, are a free form in the sense that the system does not define which ones or how many should be used to annotate a data source. What it does define is how they must be named when used. An endpoint, called point in Haystack, can be a sensor, command to an actuator (cmd), or a set point (sp) that can be written. Basic types in Haystack include points, equipment, and site. For example, a point which is a temperature sensor may have some or all of the following tags, as well as others, associated with it:

```
temp, sensor, unit: 0F, air, discharge, equipRef:…, siteteRef:…
```

All of these are Haystack tag names that, if used, must be named as specified in the appropriate tag library. Most of the names are self-explanatory except perhaps for equipRef and siteRef which are relationship tags used to denote associations among the entities, in this example with equipment and site, followed by their respective names.

The Haystack origins are evident from their entity and tag definitions that, while numbering over 200, are mostly related to building automation, lighting, and energy management [9]. However, the Haystack architectural approach to metadata handling is applicable to other domains and IoT in general. It may be used to influence and inform other specifications and conceivably be expanded through its crowdsourcing approach to other domains.

The Haystack philosophy is that system designers/installers have the latitude to annotate as much as they deem useful, and applications use annotations to deduce context and interpret tags that are meaningful to them and are programmed to process.

Haystack tags are crowdsourced and community defined in the sense that they usually start by an interested group of experts who propose definitions for comments and adoption by the relevant Haystack working group. Such tags may be proposed to the community for comments and adoption into the specification. This form of crowdsourcing is the method used by Haystack in defining tags. Designers of systems with things for which no suitable tags exist are encouraged to define their own in following the general rules and style of Haystack.

In addition to handling metadata, Haystack also specifies some operations and REST APIs for querying and retrieval of data and tags by id or filters and for subscribing and unsubscribing to notifications on watched items. However, it is not intended to function as a complete information model for constrained things as some other specifications described in this chapter.

Data serialization of Haystack payloads is supported in JSON, CSV (comma-separated values), Zinc (CSV-like with some data typing), and grids (two-dimensional tabular data structures).

The Project Haystack originally started in 2011 as a definition of a collection of tags for building management and automation. It has since gone through several iterations that added APIs for the RESTful access and, in its latest incarnation, elements of taxonomy and ontology. The taxonomy is basically a hierarchical ordering with subtyping that reflects inherent structural relationships between entities and tags. An ontology provides a more dynamic way to represent relationships in a given incarnation of a system, such as its energy or liquid flows. There is an effort under way to map Haystack types into an RDF Schema class model. Use of an ontology is expected to help model more complex system-wide flows such as electrical (relationship of meters, submeters, and loads), air (relationship of HVAC zones, AHUs, and VAVs), and water and steam distribution.

As per their official documents, the vision of Project Haystack is to streamline the use of data from the Internet of Things including, but not limited to, building and energy systems, by creating a standardized approach to defining "data semantics"-related services and APIs to consume and share the data and its semantic descriptors. The Project Haystack makes it easier to unlock value from the vast quantity of data being generated by smart devices by making the data "self-describing."

Temperature Sensor in Haystack

Haystack does not explicitly define a temperature object. Instead, it provides tags that may be associated with a temperature point. Table 7.3 contains a subset of Haystack tags that are used as example in Fig. 7.2, in the order of their appearance.

Figure 7.2 illustrates a possible representation of a water temperature sensor in Haystack notation.

This example depicts a water temperature sensor that is reporting values in degrees F, whose current value is 42.3. It uses JSON-like notation to express name-value pairs where names are Haystack-defined tags. In sequence, "id" tag denotes a user or a system-assigned unique identifier of the associated entity. The "dis" tag stands for display. It is meant to be used for human consumption, and its value is not interpreted by the system. For example, it may be used as the label to denote the associated entity on the dashboard. The "m:" value in sensor, temp, and water tags stands for "marker" which is Haystack designation for subtyping metadata. The tags "minVal" and "maxVal" indicate minimum and maximum permissible values for this entity. The last item is the time stamp of the reading. Time stamps are commonly included in sensor reports, but their definition is surprisingly omitted by many IoT standard specifications.

Table 7.3 Haystack sample tags [9]

Tag	Description
id	Unique identifier for an entity
sensor	Classifies a point as an input, analog/binary signal, or sensor
temp	Temperature measured in °C or °F
water	Point associated with the measurement or control of water
unit	Unit of measurement identifier from unit database
curVal	Current value of a point or other value records
minVal	Applied to point to define the minimum value to read from a sensor or to write from a command/set point
maxVal	Applied to point to define the maximum value to read from a sensor or to write from a command/set point
DateTime	An ISO 8601 time stamp followed by time zone name

```
{
    "id": "49cb050b77ba4500844a3fa97c2ee3ad",
    "dis": "r:ghay.ahul.cwt",
    "sensor": "m:",
    "temp": "m:",
    "water": "m:",
    "unit": "F",
    "curVal": "n:42.3",
    "minVal": "n:34",
    "maxVal": "n:45",
    "DateTime": "t:2019-07-15 14:50:07Z"
}
```

Fig. 7.2 Temperature sensor report in Haystack representation

As indicated, Haystack tags can express additional attributes that may be of interest to applications, such that it is a water temperature. They could also express additional items of interest, for example, that it is part of a particular air-handling unit and operates in a specified HVAC zone on a given schedule. Most other specifications have difficulty expressing such attributes, so those tags are not included here to make for a more even comparison.

OCF (Open Computing Foundation)

OCF (Open Connectivity Foundation) is an industry group whose stated objective is to provide a framework for secure connectivity and interoperability of IoT devices across OSs, platforms, modes of communication, transports, and use cases [10]. The group claims over 300 members, including Intel, Cisco, Qualcomm, Samsung,

Microsoft, Electrolux, LG, and Haier. The group was formed in 2014 under the name OIC (Open Interconnectivity Consortium). It subsequently expanded by merging with several competing and complementary efforts, such as AllJoyn (previously called AllSeen) and UPnP, and changed the name to OCF.

OCF specifies a complete IoT framework that includes information model (resource and interaction model), identification and addressing, discovery, messaging, device management, and security [11]. It also provides bridging specifications and translations for interactions with non-OCF devices, such as AllJoyn, oneM2M, Bluetooth, and ZigBee.

In order to communicate and interoperate, compliant nodes must implement relevant parts of the framework specification. The endpoints of each interaction are referred to as client and server, with the server providing an instance of the representation of object of interest and the client issuing requests to retrieve a server's state or to initiate actuation.

The OCF endpoint discovery is based on CoAP. It works through advertisements or directory. A discovered endpoint can be queried for capabilities and functional interfaces that describe its resource types, methods, device configuration, and management. Directory-based discovery is also supported in configurations with directory servers.

Device management provides for monitoring and maintenance. Provisioning of devices includes onboarding to enable them to join a local network and to obtain security credentials, as well as to access OCF services and to configure settings such as geographic location and time zone.

Security provides specifications for authentication and authorization of endpoints wishing to engage in communication.

Our focus in this chapter is primarily on the data modeling aspects of IoT standards and specifications. In OCF terminology, entities are physical objects that are modeled as resources. OCF provides a resource model and an interaction model. Interactions are defined using the CRUDN (create, read, update, delete, notify) model mapped to CoAP. For each defined resource type, an OCF specification describes how it behaves in response to each possible interaction.

Each resource can have multiple properties. In addition, each resource has two mandatory properties which are link attributes "if" (interface types) and "rt" (resource types) whose variants are described in the specification. Other properties represent data and metadata and are expressed as name-value pairs. For example, sensor resource has "value," "range," and "units" properties. Property named "links" can be used to link a resource to others and create collections.

Resource-type property is a reference to a machine description. As described later, those are available in machine-readable form from online tools provided by OCF. The resource-type "rt" property points to resource descriptions, basically OCF models of physical objects. In OCF resources are described using open API specifications and JSON schema. APIs define allowable methods and interface types that specific resources support. JSON schema defines the payloads, properties, and data types. The property interface type, "if," defines methods that a resource type supports and payloads that they use. A resource instance may expose more than one interface type which in turn can have different methods and payloads. This may be used by the constrained devices to pick the matching subset that they can support.

A recent OCF resource-type specification [12] includes over 120 types of resources. They model physical entities, such as a variety of sensors, home automation, security, entertainment and health devices, energy measurement, and control. Specific types include airflow, battery, binary switch, brightness, color chroma and RGB, dimming, door, energy consumption and usage, humidity, ice maker, lock, refrigeration, temperature, carbon monoxide and carbon dioxide sensors, glass break sensors, motion sensor, heart rate zone, and others. OCF has stated an intent to expand their resource-type definitions to automotive and industrial domains.

OCF provides quite extensive support for developers in terms of a formal machine-readable specification, a tool for proposing new resource types (thing and object model definitions), and open-source implementations of OCF-compliant client and server code.

The OCF online tool for creation of data models and derivatives is called oneIoTa. It contains machine-readable definitions of all approved OCF resources using OpenAPI Specification (OAS) language – it formerly used Swagger – to describe interactions and JSON to describe data schema and payloads [13]. Users are encouraged to create their own models or derivatives and propose new ones to the community. This is a form of crowdsourcing where both pending and approved definitions are available for browsing.

An open-source project called IoTivity provides a reference implementation of OCF framework [14]. The project is supported by the OCF and Linux Foundation.

The OCF association also provides a compliance certification program for products that are conducted by OCF-authorized test labs.

Temperature Sensor in OCF

Table 7.4 contains the OCF definition of a temperature resource (Thing Description) taken from a recent specification [12]. Like other resource definitions in OCF, it contains a textual description of the nature and semantics of the resource for human reference, followed by the list of mandatory and optional properties of the resource.

Temperature

This resource describes a sensed or actuated temperature value. The property "temperature" describes the current value measured. The property "units" is a single value that is one of "C," "F," or "K." It provides the unit of measurement for the "temperature" value. It is a read-only value that is provided by the server. If the "units" property is missing, the default is Celsius [C]. When the property "range" is omitted, the default is +/− MAXINT. A client can specify the units for the requested temperature by use of a query parameter. If no query parameter is provided, the server provides its default measure or set value. It is recommended to return always the units property in the result.

Table 7.4 OCF temperature resource definition [12]

Property name	Value type	Mandatory	Access mode	Description
If	Array of string	No	Read-only	OCF interface set supported by this resource
Units	String	No	Read-write	The unit for the conveyed temperature value
Temperature	Number	Yes	Read-write	The current temperature setting or measurement
Rt	Array of string	No	Read-only	The resource type
Step	Integer or number	No	Read-write	Step value across the defined range
Precision	Number	No	Read-write	Accuracy granularity of the exposed value
Range	Array of integers or numbers	No	Read-write	Range of input values, specified as two-element array
Id	String	No	Read-write	OCF defined unique identifier of the resource
n	String	No	Read-write	(Name) OCF defined human understandable name of the resource

The textual preamble describes the semantics of the resource type in humanly understandable terms for developers to interpret and use in their implementations of resource instances and interactions with them. The table that follows is an example of the structure of temperature property definition in the OpenAPI OCF Specification file. It indicates the type of responses that OCF clients may get by querying an instance of a temperature resource residing at a server. Depending on the schema that a particular type of the queried interface supports, responses include mandatory property fields and some combination of the optional ones. The latter might include properties such as precision, range, and step. The resource-type "rt" field would include a reference to the OCF resource-type designation and schema, such as "oic.r.temperature."

Table 7.4 indicates that the only mandatory property that must be returned in response to a query issued to an instance of a temperature resource is the value (sensor reading). A unit designation may be optionally returned from the choice defined by the enumeration of C, F, and K as specified in the schema. If it is not, the default assumption is C, for degrees Celsius. A somewhat uncommon aspect of the OCF temperature model is that it can be both readable and writable. Reading obtains a value from the temperature sensor. Writing, in instances when supported, is used to set the desired value of the temperature, i.e., to effectively update a set point. Many other data models treat a temperature object type as read-only and use a separate object type for set points.

Figure 7.3 shows a possible format of temperature report in OCF using JSON serialization. It starts by identifying the resource type as "oic.r.temperature,"

```
{
"rt": ["oic.r.temperature"],
"if": ["oic.if.a", "oic.if.baseline"],
    "id": "60e69140aee04513af74f720751321ec",
    "n": "gocf.ahu1.cwt",
    "temperature": "42.3",
    "units": "F",
    "range": [34, 45]
    }
```

Fig. 7.3 Temperature sensor report in OCF representation

followed by an optional indication of the types of interfaces that it supports for interaction. This is followed by the unique ID of that specific instance of the temperature sensor and its humanly readable name. A temperature value is the only mandatory property, followed by the designation of units of measurement which would be assumed to be C if not specified otherwise.

Web of Things (WoT)

The Web of Things (WoT) working group [15] within the World Wide Web Consortium (W3C) produced specifications intended to enable interoperability across IoT platforms and domains. It does not specify a particular information model. Instead, the specifications provide mechanisms to formally describe IoT interfaces that allow IoT devices and services to communicate with each other, independent of underlying implementations and across multiple networking protocols.

The WoT work and specifications follow the spirit of web design, and they were originally positioned as being the "HTML of IoT things." This is somewhat of a misnomer, since HTML is a markup language that does not describe semantics of the data. It does not need to, since it basically marks up the text whose semantics are provided by the natural language in which it is written. IoT data need semantics to indicate what they represent, such as a temperature reading or units of measure. To cope with that additional requirement, WoT follows the semantic web approach of machine-interpretable formal descriptions and tooling including the Resource Description Framework (RDF) and linked data. This allows for unambiguous definitions and facilitates machine validation and implementation.

The WoT working group was formally constituted in 2016 and published a set of specifications in 2019. They include two normative specifications (WoT architecture [16] and WoT Thing Description [17]) and three informative specifications (WoT Scripting API, WoT Binding Templates, and WoT Security and Privacy Considerations).

The WoT Thing Description (TD) is the central building block in W3C WoT. It is a machine-readable representation of capabilities for a specific instance of a physical thing and ways to interact with it. Conceptually, a TD instance is an entry point for an IoT Thing Description, and it acts like the index.html of a web site. A TD instance has four main components: textual metadata about the thing itself, a set of interaction capabilities that indicate how the thing can be used, schemas for data to be exchanged with the thing in machine-readable form, and web links to express any formal or informal relations to other things or documents about it on the web. A TD structure also contains metadata describing public security configuration of the thing and bindings for protocols that the thing supports.

The general structure of a WoT Thing Description is illustrated in Table 7.5. As shown, it consists of the general metadata, interaction affordances, public security configuration, and protocol bindings.

General metadata is used to describe the data and interaction models that the thing exposes to applications. It is based on a small vocabulary that includes identity and type of the thing. Since WoT and TD in particular are intended to support interoperability across specifications, it does not define any specific object types of things, such as a sensor or a switch. It relies on other vocabularies and ontologies, such as SSN [18], SAREF [19], QUDT [20], or iotschema [21], to do that. Thus, unlike other specifications, the type of TD does not define object types but points to the appropriate semantic definition of those, such as `"@type"`: `"saref:LightSwitch"`, where the SAREF definition of a light switch is used. For this to work, target specifications need to be machine-readable and understandable ontologies or vocabularies.

WoT specifications refer to the listing of possible interactions as interaction affordances. The term affordance comes from the human-computer interaction milieu which in turn borrowed it from the field of psychology. It basically means all interactions (transactions) that are possible between an environment and an individual or an entity. The connotation is that the individual may not be aware of all of them, but they exist in the environment and may be discovered.

In WoT there are three types of interaction affordances: property, event, and action. Any of them can optionally specify a data schema if one is not fully specified by the protocol binding that is being used.

Property is an interaction affordance that exposes the state of a thing. It must be retrievable (readable) and can optionally be updatable (writable) and observable. A thing may choose to make some of properties observable, meaning that it will push their new state after a change.

Table 7.5 Structure of WoT Thing Description

WoT thing description
General metadata
Interaction affordances (property, action, event)
Public security configuration
Protocol bindings

Action is an interaction affordance that allows a client to invoke a function of the thing. Action may manipulate a state that is not directly exposed, i.e., is not a property, manipulates multiple properties at a time, or manipulates properties based on the internal logic state such as toggle. It may also invoke an action/trigger that changes state over time, such as fading of a light.

Event interaction affordance describes event sources that push data asynchronously from the thing to the consumer. In principle, an event is a transition of a state, as opposed to a state which is a property. Events may be triggered by conditions that are not exposed as properties, such as alarms. The WoT also considers as events individual readings in a time series that are pushed at regular intervals.

A Thing Description describes the metadata and interfaces of things, where a thing is an abstraction of a physical entity that provides interactions to and participates in the Web of Things. Thing Descriptions provide a set of interactions based on a small vocabulary that makes it possible both to integrate diverse devices and to allow diverse applications to interoperate. Thing Descriptions are encoded in JSON-LD. This allows the TD to be processed as a JSON document.

In WoT systems, virtual entities can also be represented as things with thing descriptions. A Thing Description instance can be hosted by the thing itself or hosted externally for legacy and constrained devices.

When describing implementation, WoT specifications refer to clients as consumers and introduce the notion of servient modules that can act as a client or a server, or both. For example, a servient residing at an intermediate point, such as a gateway, can act as a consumer (client) of the thing on the local network and as a thing, via its TD proxy, to consumers located elsewhere in the system. The WoT architecture supports intermediaries and proxying via digital twins in order to handle intermittent connectivity and network topologies that include firewalls and NAT (Network Address Translation) units in order to deal with security and address limitations in IPv4 systems. In fact, the architecture is quite flexible in terms of comprehending different system topologies and types of connections, including device-device, device-gateway, gateway-cloud, and even inter-cloud connections in federated systems.

In complex compositions, TDs may link to (hierarchical) sub-things in the composition. Linking is also used to express relations between things and other resources in general, e.g., online manuals or design documents. Web linking among things makes WoT navigable for both humans and machines. This can be further facilitated by providing thing directories that maintain catalogs of available things.

Protocol binding is the mapping from an interaction affordance to concrete messages of a specific protocol, such as HTTP, CoAP, or MQTT. All exchanged data must be identified by the media type, e.g., application/json or application/cbor or application/ld + json.

Security metadata in TD describe the type of public security configuration that the thing exposes and that the prospective consumers need to adhere to in order to interact with the thing.

OPC (Open Platform Communications)

Open Platform Communications (OPC) is a factory automation standard. It claims over 450 members and over 1500 compliant products [22]. It is one of the leading legacy systems used in control automation that IoT systems may expect to interface with for acquiring sensor and machine-floor status information.

OPC started in 1995 with the focus on providing a simple open standard for the first-tier visualization and for applications to be able to read-write and subscribe to data from factory automation. It originally used the Microsoft proprietary and now largely deprecated COM (Common Object Model) and the DCOM (Distributed Common Object Model) object technology. Trying to avoid the COM-DCOM limitations, OPC created an XML-based data access version, XML-DA, whose implementations turned out to be slow. The latest version, the OPC Unified Architecture (UA), aims to embrace Internet technologies and position itself for the Internet of Things. It also has a requirement to retain some backward compatibility, which results in a somewhat strained amalgam of the Internet and distributed systems technology of the past.

The OPC UA has an elaborate specification consisting of 13 parts that include security model, address space model, information model, services and service mappings, alarms and conditions, discovery, and aggregates [23].

Even though it includes specification of the information model, the OPC UA does not actually define one. Specifics of the information model for particular products and things are left to manufacturers and industry associations. However, the OPC UA specification does provide a formal structure for representing the information model using objects, properties, references, and methods. It also formalizes how clients (services and apps) can browse this information on the server and interact with the modeled objects. Moreover, the linking mechanism provided by object references allows modeling of complex physical installations, such as production lines or entire plants.

The OPC UA Information Model uses the concept of an object to represent real-world entities. Objects can represent simple entities, such as a temperature sensor or complex linked assemblies. Information models are represented as highly structured XML files, stored in the OPC UA address space with defined mechanisms on how to peruse them. Objects may include component objects, variables, methods, and references to other objects. A rich set of references include Organizes (folders containing collections), hasTypeDefintion, hasComponent, and hasProperty. Variables contain data. Methods are programmed actions available to clients to access prescribed manipulations.

Servers have address spaces organized in a structured manner that clients can access and browse for object definitions, system hierarchy detection, reading values, and the like. The address space contains object definitions, simple or complex linkages that can represent a set of machinery, product line, or even the entire plant. The address space starts with a root node that links to objects below that can be folders (directories) of objects, types, views, etc. Clients can browse server address spaces, i.e., traverse the hierarchy starting from the root. This is done by accessing the OPC UA stack on the server. Servers also have a separate communications stack that is used to identify communications attributes like protocols and security and to establish client-server communications channels and sessions.

Figure 7.4 illustrates the architecture of an OPC UA server. As indicated, it includes a collection of nodes that are software representations of real objects and physical things such as sensors or counters. Clients can access server functions and its address space through OPC UA server APIs. A server implementation can choose to make some nodes in its address space visible and browsable by clients to read their state and properties and actuate outputs. Nodes can be linked in various ways, such as hierarchies and meshes, to reflect their connections and relations in the real world.

Servers can also create monitored items that notify clients when defined events, such as changes of state or an alarming condition, occur. Clients can regulate the frequency of notifications in their subscriptions to published messages.

The OPC UA supports many-many connections between clients and servers and server-server communications in peer-to-peer fashion.

The OPC UA also includes specification of security and discovery. In addition, it covers alarms and event management that are commonly found in implementations of industrial automation. The specification is not dependent on a specific communication protocol. It supports the UA TCP that works with smaller nodes using binary encodings, but it can also work with HTTP and HTTPS. Data and payload serializa-

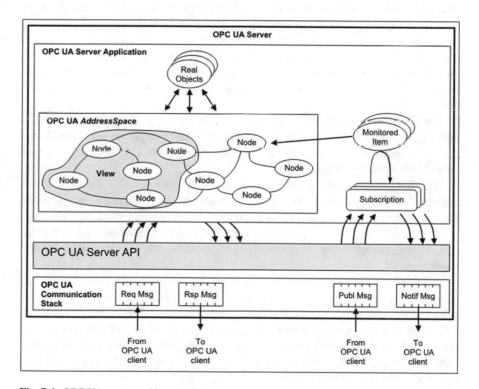

Fig. 7.4 OPC UA server architecture [24]

tion options include XML, SOAP/XML, and OPC UA binary. Servers also have a separate communications stack that is used to identify communications attributes like protocols and security and to establish client-server communications channels and sessions [25, 26].

An OPC UA server can reside on an edge device and handle its physical I/O and client interactions and gives notifications. OPC UA defines several server profiles of different complexity that indicates what the server can do. Clients can issue requests to servers that are supported by the server profile and implementation, such as read an attribute of a node such as temp on channel #2, browse address space, add or delete nodes, subscribe to notifications, and request historical data or diagnostics.

Server objects, located in the client-accessible address space, can provide additional information, such as other servers that contain namespaces from other organizations used by the local objects, server operational status, the manufacturer, OS and software builds and versions, and product codes. Servers can also keep track of and report their quality of service and reliability for considerations in redundant server configurations and indicate if they are auditing events.

The OPC UA supports server discovery. This is supported by a three-tiered SOA-like mechanism consisting of Local Discovery Servers (LDS), Local Discovery Server with Multicast Extensions (LDS-ME), and Global Discovery Servers (GDS). Clients discover servers using the OPC UA discovery mechanism. Once a server of interest is selected, a client accesses it via Get Endpoints request that returns server security requirements, public credentials, and access points for the supported protocols, such as binary or HTTPS. The client then chooses the access point corresponding to the transport of interest, such as binary or HTTPS, and uses it to establish a specific secure communication channel and session with the server. Long-lasting channels are used for client-server communications, and sessions are used between applications.

The default client-server communication mode is request-reply, but OPC UA also supports a publish-subscribe mechanism with and without brokers. Details are not well specified and are mostly left to the implementer, thus creating potential cross-vendor interoperability issues.

The OPC UA security is designed to guard against security threats deemed important to factory automation settings, including message flooding, unauthorized disclosure of message data, message spoofing, message alteration, and message replay. Security mechanisms include certificate-based authentication and authorization, confidentiality using RSA and SHA algorithms, and integrity using digital signatures.

The OPC UA defines three types of access to historical data for variables, events, and address spaces. If history is available for a variable, clients can query its past values within a supported range. History of events is less well specified, but the idea is to allow clients to obtain history of events – if maintained – within a certain time range, such as during the past hour. Address space history refers to tracking and reporting node configurations in the address space, an optional feature peculiar to the OPC's address space structuring and management.

Closing Remarks and Observations

There are a number of standards and industry specifications that provide IoT information models and frameworks to a different extent. They have similarities and differences, many of which are illustrated by the representatives described in this chapter. They all model physical objects, such as sensors and actuators, into software representation that servers who are hosting instances of the resources can use to reflect and report their states. Consuming applications and services, clients use the data models to create their own instances of thing representations and to interpret responses from servers. Use of the common specification results in syntactic and semantic interoperability in the sense of interpreting what the data means.

In principle, specifications are intended to be used at the design time for creating thing representations that can meaningfully interact and interoperate in exchanges between compliant IoT endpoints. The server and client side may be developed independently and on different platforms with the expectation of interoperability by virtue of adhering to the common specification – just like the Internet.

IoT standards strive for and often claim device or cross-platform interoperability. In most cases this means interoperability among devices that use the same specification and in effect provides intra-specification interoperability. As discussed in chapter "IoT Data Models and Metadata", other levels of interoperability across domains and standards are valuable and may need to be implemented by building aggregation and translation layers that operate above the individual specifications.

The presented specifications have many similarities that result from the fact that they are modeling the same set of real-world physical things and from the common structural approach of using software abstractions inspired by the object-oriented approach. Semantic interoperability is mostly accomplished by defining the controlled vocabularies and structure of names and properties of object types. The size of the vocabulary corresponds to the number of things that are defined by a given specification. Their types and nature of objects reflect and in effect determine the primary application domain that a specification can be used for, such as consumer, industrial, transportation, health, and buildings.

Most specifications use the name-value pairs to represent attributes and properties of objects, including metadata. This lends itself to the common use of JSON and JSON-LD for serialization.

Differences, aside from the terminology issues discussed earlier, are primarily in the choice and coverage of the specified types of objects and in the ways that they are modeled. In addition to the obvious differences in naming and syntactical representation, the specifications sometimes differ in selections and clustering of properties when modeling the same objects, such as temperature sensor.

Issues related to some specifications include restrictiveness of prescriptive models, limitations in handling metadata, and relatively small numbers of defined object types.

Restrictiveness can result from the use of an object-oriented approach with a tendency to fix properties that may be associated with a given object. A common problem with this approach is the inability to add metadata annotations that are not included in the specified list of object attributes but may be useful to applications.

Other problems can arise from the incongruence between the nature of the physical thing and its model. For example, one standard allows only Boolean values to express the output of a CO2 sensor signifying that CO2 is present or not present. However, many CO2 sensors report their readings in ppm (parts per million). This means that the developer of a sensor's software handler is forced to substitute some potentially arbitrary metrics for translating the numeric readings into the two values allowed by the model. Moreover, the actual readings in ppm provided by the sensor cannot be made available to the processing applications and storage.

In principle, the solution is to make specifications more descriptive and less prescriptive. In other words, specify how to name and use a property when needed, but do not restrict which properties can be associated with an object. The Internet solved this problem elegantly a long time ago – browsers interpret the tags that they understand and ignore the ones that they do not. The Haystack specification provides a workable example of how to do this with metadata.

Probably the most significant problem is the relatively small number of object types in each specification. The number of object-type definitions in specifications tends to be relatively low, below, or not much above 100. This is mostly due to the amount of time it takes to reach consensus in voluntary deliberative bodes, like industry standards organizations. One of the consequences is the slow growth of the formally agreed-upon and specified types of objects. The problem is that there are orders of magnitude more things of interest to IoT systems that have yet to be modeled, and their number is growing faster than the specifications can cope. This leaves system implementers in a dilemma as to what to do when their systems include things for which object types are not specified. In the long run, this problem may be ameliorated by the use of formal taxonomies and ontologies that are being worked on.

In the meantime, if the intent is to stay with standards, probably the best course of action is to pick a specification that is the closest approximation to system requirements and design preferences and use it for the object types that it specifies. Object types that are part of the IoT system but not formally specified should be defined by closely following the architecture and the design style of the chosen specification. This usually helps to reduce the amount of subsequent changes. In addition, that new object type may be proposed for adoption. Several standards bodies and industry groups encourage and provide tools for such crowdsourcing style of submissions.

References

1. OMA SpecWorks [Online] Available at: https://www.omaspecworks.org/ (Accessed Dec 15, 2019)
2. Lightweight M2M [Online] Available at: http://www.openmobilealliance.org/release/ LightweightM2M/Lightweight_Machine_to_Machine-v1_1-OMASpecworks.pdf (Accessed Dec 15, 2019)
3. Jimenez, Koster and Tschofenig (2018) *'IPSO smart objects'* [Online] Available at: https:// www.omaspecworks.org/wp-content/uploads/2018/03/ipso-paper.pdf (Accessed Dec 15, 2019)
4. Nottingham, M. 'Web linking' (2017) *IETF RFC 8288* [Online] Available at: https://tools.ietf. org/html/rfc8288 (Accessed Dec 15, 2019)
5. Shelby, Z. (2011) 'CoRE link format' *IETF Internet-Draft* [Online] Available at: https://tools. ietf.org/id/draft-ietf-core-link-format-02.html (Accessed Dec 15, 2019)
6. IPSO objects in OMA LwM2M object and resource registry [Online] Available at: http://open-mobilealliance.org/wp/OMNA/LwM2M/LwM2MRegistry.html (Accessed Aug 15, 2019)
7. Project Haystack [Online] Available at: https://project-haystack.org/ (Accessed Aug 15, 2019)
8. Haystack Developer Site [Online] Available at: https://project-haystack.dev/ (Accessed Dec 15, 2019)
9. Haystack Tags [Online] Available at: https://project-haystack.org/tag (Accessed Dec 15, 2019)
10. Open Connectivity Foundation [Online] Available at: https://openconnectivity.org/ (Accessed Dec 15, 2019)
11. OCF Specifications [Online] Available at: https://openconnectivity.org/developer/specifica-tions/ (Accessed Dec 15, 2019)
12. OCF Resource Type Specification 2.0.2 (2019) [Online] Available at: https://openconnectivity. org/specs/OCF_Resource_Type_Specification_v2.0.2.pdf (Accessed Dec 15, 2019)
13. oneIoTa [Online] Available at: https://www.oneiota.org/ (Accessed Dec 15, 2019)
14. IoTivity [Online] Available at: https://iotivity.org/ (Accessed Dec 15, 2019)
15. W3C Web of Things Working Group [Online] Available at: https://www.w3.org/WoT/WG/ (Accessed Dec 15, 2019)
16. Web of Things Architecture (2019) [Online] Available at: https://www.w3.org/TR/wot-archi-tecture/ (Accessed Dec 15, 2019)
17. Web of Things Thing Description (2019) [Online] Available at: https://www.w3.org/TR/wot-thing-description/ (Accessed Dec 15, 2019)
18. Semantic Sensor Network Ontology (2017) [Online] Available at: https://www.w3.org/TR/ vocab-ssn/ (Accessed Dec 15, 2019)
19. ETSI TS 103 264 Smart M2M; Smart appliances; Reference ontology and oneM2M mapping (2017) [Online] Available at: https://www.etsi.org/deliver/etsi_ts/103200_103299/103264/02. 01.01_60/ts_103264v020101p.pdf (ASccessed Dec 15, 2019)
20. QUDT [Online] Available at: http://www.qudt.org/ (Accessed Dec 15, 2019)
21. iotschema [Online] Available at: http://iotschema.org/ (Accessed Dec 15, 2019)
22. Open Platform Communication [Online] Available at: https://opcfoundation.org/ (Accessed Dec 15, 2019)
23. OPC Unified Architecture Specification [Online] Available at: https://opcfoundation.org/ developer-tools/specifications-unified-architecture (Accessed Dec 15, 2019)
24. OPC Unified Architecture Specification, Part 1: Overview and concepts [Online] Available at: https://opcfoundation.org/developer-tools/specifications-unified-architecture/part-1-over-view-and-concepts/ (Accessed Dec 15, 2019)
25. Mahnke, W et al., (2009) *OPC Unified Architecture*. Berlin: Springer Verlag.
26. Rinaldi, J. S. (2018) *OPC UA unified architecture: the everyman's guide to OPC UA"*, San Bernardino, CA: Rinaldi Publishing.

Chapter 8
IoT Platforms

An IoT platform is a collection of integrated components that provide much of the common functionality required to implement an IoT system. IoT platforms have the potential to reduce complexity, cost, and length of product development by providing and supporting many common core system functions in a professional and presumably trustworthy manner. In principle, their use should enable IoT system designers to focus on their areas of expertise and the product's unique added value and to rely on IoT platforms to complete much of the rest of the system needed for a complete system offering [1].

Design and implementation of an IoT system generally require expertise in a number of diverse areas which may not be readily available in smaller and medium-sized companies. For example, a start-up company developing a novel personal fitness device would naturally focus on things that make their offering and value proposition unique. These may include expertise and activities such as miniaturized sensor and component development and integration, battery design and management, embedded system design, some wireless connectivity, firmware, and data fusion to detect combined measurements that jointly may infer types of user activity. However, the customer-facing product may need to include an entire IoT system for data acquisition and processing, cross-user data aggregation for summary statistics, detection of cross-population trends, and user-defined friends and groups activity comparisons and competitions. In addition, a mobile phone application may be needed to provide a communications gateway between the device and the Internet and provide a user interface for data visualization, settings of goals, and notifications. The creation of such a system requires many additional skills, such as distributed IoT system design, networking, security, database management, scalability, server and cloud programming for gateway and cloud portions, device and user management, authentication, service and data replication for reliability and resilience, mobile application and back-end development, and the like. These skills are difficult to acquire and justify in terms of costs for all but very large organizations.

There are many variants of IoT platforms that are available from commercial vendors and the open-source community. Since the most elaborate software parts of

© Springer Nature Switzerland AG 2020
M. Milenkovic, *Internet of Things: Concepts and System Design*,
https://doi.org/10.1007/978-3-030-41346-0_8

a complex IoT systems commonly involve components in the cloud, commercial cloud providers tend to provide rather comprehensive and robust IoT platforms that build upon their existing infrastructure and technology. In this chapter we describe in some detail two of the more popular ones, Amazon AWS IoT and Microsoft Azure IoT platforms. The choice was based primarily on the breadth of coverage, relevance and market presence, and not on a vendor preference. Our intent is to illustrate the scope, coverage, and common characteristics of IoT platforms that can be expected from commercial providers. The reader is referred to the vendor literature for details of particular platforms.

IoT Cloud Platforms

In general, an IoT platform needs to facilitate the implementation and integration of required end-to-end data and control flows and processing that leads to insights and actions on the IoT system. In addition, it needs to be reliable, customizable, and support the required levels of scalability, extensibility, and resilience. IoT platforms often include additional features such as implementation of digital twins and facilities for flexible placement of processing functions along the data route, such as analytics at the edge. In this section we summarize some of their common characteristics and describe specific implementations in the subsequent ones.

Depending on the scope, an IoT platform may include software, hardware, development tools, and integration frameworks that facilitate some or all of the following:

- Data acquisition from the edge
- Data ingestion and routing based on rules and message content
- Integration with cloud services
- Security management and enforcement
- Device management
- Cloud services, such as complex event processing, notification, visualization, storage, business intelligence, and analytics
- Development tools and kits (SDKs)

In addition, an IoT platform needs to be highly scalable, support millions and even billions of endpoints, reliable, and professionally managed for security and high availability.

The IoT-specific parts of the commercial cloud platforms are generally centered around a cloud-based IoT core or hub that establishes connectivity and security between edge devices and cloud services. It is typically designed and implemented by the platform provider and operated as a managed service. An IoT core typically consists of a gateway that implements a bidirectional messaging interface to external IoT endpoints. Incoming messages are filtered and routed to their intended functions and services in the cloud, such as the stream processing and storage. The gateway is a routing and connection and integration point for services and functions available from the cloud provider. Some of them may be specifically designed for IoT uses; others are more general Internet services that may be applicable. In addition, users can deploy their own processing functions in the cloud and have the gateway route messages to them as designated by the filtering rules in effect.

Since the IoT core is hosted in the cloud, it is commonly accessed over an Internet-compliant public or private network using standard protocols such as TCP, HTTP and in many cases MQTT. The endpoints need to be able to support this mode of interaction, either directly in the case of smart things or indirectly via the services of the edge gateway.

The gateway also serves as a security boundary between the cloud and edge devices. In order to assure compliance with the usually stricter levels of security required within the cloud, the IoT core is usually in charge of authenticating edge devices on an individual level and authorizing specific types of exchanges to individual actors. In this way, the IoT core can enforce required security practices, such as certificate-based authentication and revocation of credentials from compromised devices. Communication with endpoints is typically encrypted at the transport layer and above. Security credentials are issued and validated as part of the provisioning process, before an endpoint is allowed to join the system. Device management functions include provisioning and readying of devices to join the system and their state and operational monitoring thereafter. They are also in charge of performing software and security updating and patching.

A rather common part of IoT platform offerings include external edge gateway runtimes that may be installed in nodes outside of the cloud. They are designed to host some preprocessing functions locally, support the cloud core connection protocols and procedures, and sometimes provide the local autonomy and integrity in the form of supporting disconnected operation. Some vendors also provide a customized operating system for smaller microcontrollers and SoCs common in smart things and edge devices with libraries to facilitate implementation of cloud connectivity and security.

IoT core and vendor provided cloud services are typically offered as managed services. That means that the vendor is responsible for providing the code, execution environment and instantiating the service, scaling it as necessary, and maintaining its reliability and resilience. It also means that the user is responsible for paying for it, typically on a usage basis. IoT platform vendors also provide the tools and system development kits for creating applications for their system and in some cases for porting the edge software to the supported hardware platforms, such as things and gateways.

Due to the immense variety of sensor types and IoT things at the edge, IoT platforms generally do not provide direct support for arbitrary raw sensor connectivity. This means that users are responsible for procuring software that interfaces to their specific sensors, provides the necessary conditioning, and brings that data to a processing point that can be converted into the format suitable for sending to the cloud. Some organizations are working on creating a sensor equivalent of plug and play that worked quite well for the PC ecosystem.

Amazon Web Services (AWS) IoT Platform

Amazon has an IoT platform offering that works in conjunction with their other managed cloud services [2]. The IoT-specific part is referred to as the AWS IoT. It consists of a cloud-hosted set of integrated functions called the IoT core. The offering also includes an edge gateway software base for external execution called Greengrass and FreeRTOS real-time kernel for microcontroller-based things. Both provide mechanisms and protocols for connecting to the IoT core.

AWS IoT Core

Key components of the AWS IoT core are depicted in Fig. 8.1. It consists of a data plane and control plane. The data plane is in charge of getting sensor data into the cloud for processing and storage. It consists of a message broker, rules engine, and device shadows (AWS name for digital twins).

As indicated in Fig. 8.1, the key core components include:

- Device connectivity and data ingestion
- Data routing and service integration, rules engine
- Devices shadows (digital twins)
- Control plane: security, authentication, and management

Data Plane

The message broker supports pub/sub via MQTT and allows sensors and edge devices to securely publish their reports to topics established for that purpose. It instantiates MQTT brokers that forward published messages to active subscribers. An AWS IoT message broker supports the native MQTT protocol and a web socket mode for access from browsers and mobile devices. HTTP is supported for publish-only endpoints.

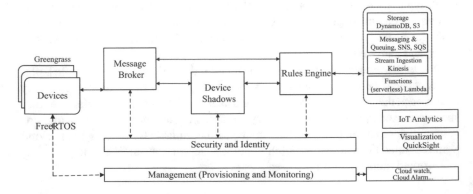

Fig. 8.1 Amazon Web Services (AWS) IoT platform

Applications and use cases that do not need messaging can use the basic ingest mode that does not incur messaging costs. It works by using the topic names that are parsed by the rules engine to trigger actions and route data to their target processing functions.

Posted message payloads are processed by the rules engine that routes them to other services, such as storage and functional modules in the AWS Lambda that implements a form of a serverless system.

The IoT core implements a mode of digital twins that Amazon literature refers to as device shadows. Shadow is a persistent digital representation of a device that contains its last reported state and, in case of actuators, the desired state. Cloud services access shadows to obtain the last known reading and to initiate actions by modifying the desired state to what an output should be. If there is a discrepancy between a reported and a desired state, the shadow instructs the endpoint to initiate the necessary action. Completion of the action is confirmed when the endpoint reports a new state that matches the desired one.

The level of indirection provided by the device shadows allows applications and services to operate on the last known data even when devices are disconnected, thus minimizing disruptions caused by the temporary communications outages. Another benefit is that cloud services can operate with a simple output abstraction of the desired state, with the edge software handling device-specific steps and protocols. Device shadows are implemented as JSON documents that may be queried and updated to reflect the reported and the desired changes.

Jobs service, not shown in Fig. 8.1, is the last major component of the data and control plane in an IoT core. It allows users to define and trigger actions or set of actions to be sent and executed by one or more devices connected to the AWS IoT. These can target individual actions, such as actuation, or group of devices to perform activities such as downloading and installation of updates.

Control Plane: Security and Management

The AWS IoT control plane consists of security and management components. Security and identity service manages secure connections and payload exchanges with end devices. It authenticates endpoints and servers using X.509 certificates and Amazon authentication services such as IAM (identity and access management) and Cognito. Users can supply their own certificates signed by a certificate authority for AWS IoT to verify and use for authentication purposes. The system handles authentication of devices to AWS IoT system and services, and it allows devices to request server authentication if they want to make sure that they are communicating with authenticated AWS IoT entities.

Authenticated entities are subjected to authorization that increases security and system integrity by limiting what each entity can do. Authorizations are expressed in terms of policies that must be attached to each entity before it is allowed to access the system. The AWS IoT allows the use of custom authentication services, such as JSON Web Tokens and OAuth, provided that they can return proper policy documents.

The AWS IoT message broker and device shadow service use TLS to encrypt all communication. TLS is used to ensure the confidentiality of MQTT and HTTP application protocols. For MQTT, TLS encrypts the connection between the device and the broker. TLS client authentication is used by the AWS IoT to identify devices. For HTTP, TLS encrypts the connection between the device and the broker.

Management services include device registration, provisioning, and remote management. Devices that intend to operate as a part of an AWS IoT installation are recorded in the AWS IoT registry. Device registry entries can include custom attributes, security certificates, and MQTT client IDs. Device registry supports searches to locate devices based on the combinations of attributes, such as device ID, state, and type.

Registry entries may be organized into groups and hierarchies with parent-child relationships. This allows group-level management as any action performed on a parent group is also applied to its child groups and to all devices therein. The device provisioning service allows devices to be provisioned individually and in groups. A provisioned device or thing has a registry entry that contains its unique identifier, description, digital certificate, policy attached to the certificate, and a set of attributes of the thing, including existing types and groups. Device provisioning may be simplified by using provisioning templates. Devices can be provisioned individually, in bulk, or just-in-time when they first connect to the AWS IoT.

The AWS IoT device management supports remote management functions that include updating of software, downloading and execution of security patches, and control actions such as reboot and reset to the factory state.

The AWS IoT device defender is a security management service that can audit device security configuration and monitor connections for abnormal activities that may require remediation. Its audit function verifies that devices security policies are set in accordance with the policies in effect. It can also help to detect security drifts from best practices across multiple policies. This can happen as multiple policies are defined and revised incrementally over time, perhaps resulting in overly permissive policies that may allow some devices to read or update data from many others or in multiple devices having the same identity. Device defender can also help to identify unusual behavior, such as spikes in outbound or inbound traffic and communications with unauthorized or unrecognized endpoints that may be malevolent. The AWS IoT can monitor traffic on the cloud side. Device-side monitoring requires users to install device-side metric, connect it to the service, and create a profile to describe what constitutes normal and abnormal behaviors.

AWS Supporting Cloud Services

In addition to the services designed specifically for IoT, AWS users can make use of other suitable services in an AWS portfolio that IoT core can invoke by means of the rule engine. They include databases, stream processing, serverless functions, and notification and queuing services. Of the many databases available on AWS, Amazon Simple Storage (S3) and DynamoDB seem to be the most applicable to IoT uses.

AWS S3 is a large scalable storage for Internet uses that may be accessed through web services [3]. DynamoDB is Amazon's implementation of NoSQL database [4]. Several NoSQL databases available from other sources, such as MongoDB, are also available for IoT platform use as AWS hosted services.

Other AWS services that may be applicable to IoT uses include Kinesis, Simple Notification Service (SNS), and Simple Queue Service (SQS). Kinesis is a stream processing service originally designed for Internet uses such as processing of click streams, social media posts, and location events [5]. It has separate modules for processing of data streams, video streams, data analytics, and a "data firehose" service for capturing, transforming, and loading streaming data into storage, such as S3.

The AWS SNS service can send and receive notifications from a variety of sources, including SMS, email, and social media [6]. Amazon SQS stores data in queues that may be retrieved by applications.

AWS provides extensive facilities for monitoring of its own and customer services. They include monitoring and processing of cloud events, alarming when some or combinations of specified metrics exceed thresholds, and logging and monitoring of logs created by those subsystems. In addition to those, the AWS IoT dashboard allows manual monitoring of items not covered by the tools.

AWS IoT Services Layer

At the cloud application and services layer, Amazon has AWS IoT Analytics and the ability to interface with its general visualization service for business intelligence, called QuickSight.

AWS IoT Analytics is a combination of customary AI-tool functions with some additional provisions for the requirements of IoT systems [7]. Since IoT data can be noisy and corrupted, it facilitates data cleansing, transformation, and enrichment. This is accomplished through integration with (mostly user supplied) functions that can do useful things including data filtering, unit translation, elimination of outliers, and data interpolation where there are gaps. Data can be enriched with integration of external sources, such as weather forecasting services.

IoT Analytics supports the processing of streaming data that enables continuous monitoring and analysis of performance or anomalies over time. Input data are placed in a time-series data store that is optimized to handle queries by time, which is common in IoT systems. This is a transient message store, not a traditional database. Messages can be replayed by epochs for reprocessing.

IoT Analytics does not provide a specific ML tool, but it interoperates with Amazon general analytics and ML services. These include SageMaker and numerous frameworks and pre-trained models for recommendation, forecasting, image and video analysis, text analytics, document analysis, translation, transcription, and text-to-speech (voice) conversions. Those and user-developed or trained models may be stored for use and execution in the cloud or at the edge, such as in Greengrass as Lambda functions. Models executed at the edge can send training data and model refinements back to the cloud.

AWS IoT Edge

Amazon provides software and development tools for creating and managing devices at the edge and enabling them to participate in AWS IoT systems. They include FreeRTOS and Greengrass gateway.

FreeRTOS is a real-time operating system (RTOS) designed for microcontrollers and small microprocessors. It started as an open-source project that Amazon is now supporting with libraries for IoT and extensions for integration with AWS IoT systems [8]. It is a typical real-time kernel for embedded systems with a pre-emptive multitasking scheduler that is priority based and can support the deterministic execution of real-time tasks. Other features include several memory allocation options, including static and intertask coordination primitives such as notifications, message queues, multiple types of semaphores, and stream and message buffers. Amazon extensions include libraries for AWS IoT connectivity, security, and over-the-air updates.

FreeRTOS from Amazon is primarily intended to enable construction of IoT things and constrained devices that can be connected to AWS IoT cloud directly or via a supporting gateway, such as the Greengrass.

The AWS IoT Greengrass is software that operates at the edge, typically on user premises and outside of the AWS trusted cloud perimeter [9]. It is a form of edge gateway in that it provides for secure connection and exchange of data and commands among its locally connected devices even when disconnected from the cloud. Functionally, it provides a subset of AWS IoT core functions with the added benefits of edge processing and local autonomy. It easily connects to larger AWS IoT installations through the cloud and provides a local execution environment for cloud-developed services, such as Lambda functions and inference engines from the IoT Analytics.

In terms of data plane functions, Greengrass provides a local message broker, device shadows, Lambda runtime, and support for local execution of ML inference engines. Device shadows operate locally and may be synchronized with their cloud counterparts. Lambda functions are the AWS implementation of serverless computing. Lambda functions can be deployed from the cloud and executed on Greengrass using its local runtime. As is common in serverless approach, execution of those functions is triggered by events that may originate anywhere in the system, locally or from the cloud.

The Greengrass control plane functions include security and authentication management, support for local secrets manager, and hardware security modules including trusted platform module (TPM). Management functions include group management, over-the-air update agent, and discovery service that, among other things, enables local devices and FreeRTOS things to discover their Greengrass core on the local network.

The integration with AWS IoT cloud and local devices is supported by built-in connectors, libraries, and protocols, including OPC UA. Greengrass runs on hardware platforms with more powerful processors and operating systems, such as Linux. It can run in a container on Windows and macOS systems.

AWS Additional IoT Services

In addition to the IoT core and its functions described earlier, Amazon offers several stand-alone services apparently intended for the more specific IoT uses. They include AWS Things Graph, AWS Sitewise, and AWS IoT Events.

Things Graph is a graphical-user interface tool aimed at controlling and coordinating behaviors of a number of devices that are not designed to work together [10]. It seems to be best suited for uses like home automation. The system creates abstract device models that capture functional attributes, inputs, and outputs of the commonly used classes of devices and allows individual things to communicate with them using protocols such as the MQTT and Modbus. Things Graph provides a graphical user interface, conceptually similar to Node-RED, for creating flows that define sequences of data and control exchanges and actions among devices. In the process, it performs mapping to and from the model representations by translating between device-specific messages and their abstract model depictions.

AWS Sitewise is a service that allows collection of data from industrial equipment at scale. It interfaces to industrial and legacy protocols, such as the OPC UA, to monitor equipment for operational efficiency and faults [11]. This service basically brings a common industrial use into the cloud, with the additional potential of enabling aggregation and insights across production segments and data sources within an enterprise. It can also provide cross-facility monitoring from a central location. Cited use cases include manufacturing, food and beverage, energy, and utilities.

AWS IoT Events service is designed to streamline the steps involved in processing events at scale, including data ingestion, multicomponent event description and triggering, and building of processing rules using if-then-else logic [12]. Examples provided in its description include notification of technicians when the freezer malfunctions to prevent food spoilage and combining multiple sources of telemetry – such as belt speed, motor voltage, amperage, and noise level – to gain understanding of events such as that a motor may be stuck.

Using AWS IoT

Potential users need to make their data sources and communication modalities ready for integration and compliant with security and naming conventions of the AWS IoT. This includes interfacing sensors and things and being able to transmit their payloads using MQTT or HTTP in the basic ingestion mode. Prior to connecting, user devices and things need to be able to engage in AWS certificate-based authentication and authorization procedures and be able to define and configure policies. This may require some custom programming and implementation of AWS IoT APIs with compliant protocols and encryption methods. Amazon provides several software-development kits (SDKs) for popular platforms and languages. Use of FreeRTOS or Greengrass can speed up the process. In addition to those, prospective users need to use the AWS IoT core conventions and interfaces to deal with the publication of their topics and definition of rules to route their messages to appropriate services such as storage and visualization. This also requires development and familiarization with interfaces and usage of the chosen services, such as Lambda, DynamoDB, or Kinesis. If AWS tools and frameworks are used to complete the design of user applications and services, they may require additional development and skills.

Microsoft Azure IoT Platform

Azure IoT is Microsoft's rendering of an IoT platform [13]. Figure 8.2 illustrates its general architecture and key components [14]. As is the case with other figures in this chapter, it was constructed on the basis of vendor's documents while keeping with the general IoT core system structure outlined in chapter "Communications" to facilitate comparisons. As the figures suggest, the basic architecture and functionality of these systems are quite similar.

As Fig. 8.2 indicates, the key linchpin between the data sources and the cloud back-end processing services is a cloud hosted IoT gateway service called the IoT hub. It provides the first-point interface for data ingestion and ensures authenticated and secure communication between endpoints and cloud services. It supports brokered-based bidirectional communication and routing of messages based on their type and processing rules. The rest of the architecture is a lambda-style split into stream processing and archival storage of data for batch processing. In Microsoft terminology, these are called the warm path and the cold path, respectively. Incoming data streams are processed according to the applicable rules that can send them to functions such as aggregation and complex event detection or alerting, UI reporting, and visualization. Streaming data may also be stored in the "warm path store" for low-latency processing of recent data via simpler queries such as values by date and time range, aggregates, and single-point telemetry readings. These can generate real-time insights that may be fed to enterprise the systems via the Azure's Business Integration block to generate appropriate actions.

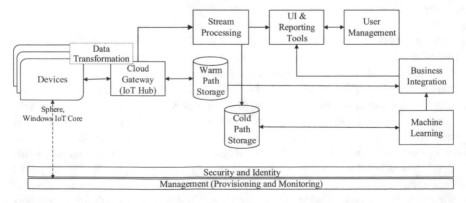

Fig. 8.2 Microsoft Azure IoT platform

The cold-store path supports more complex queries on larger aggregate data sets that may be required by analytics and machine-learning services that generally take longer time to compute.

The data-translation block indicates that some processing of data formats may be required to bridge the differences in representation between the reporting format and the aggregation and data-storage formats used by the cloud-based applications and services. Azure IoT does not enforce a particular information model, and it is left to the applications and services to define and interpret the data and metadata formats that they use.

The crosscutting control plane functions include security and management. Security is managed by performing authentication at the individual device level, rule-based authorization, and encryption of payloads at the transport and application layers. Additional security features include limiting of communication only to well-known endpoints and preventing unsolicited communications to the edge. In general, Azure IoT is architected to not trust the link layer.

Manageability deals primarily with provisioning of devices with security credentials and identity and software updates needed to comply with policies in effect in a particular Azure IoT system installation. Operational and health monitoring is provided through Azure system services.

Another crosscutting function of the Azure IoT platform cited by Microsoft literature is high availability and disaster recovery. This is a general property of hosted services in the Azure public cloud that is beneficial to IoT systems implementations but is not specifically designed for them.

At the edge and outside of cloud, the system provides an IoT Edge field gateway that is capable of temporary disconnected operation and can instantiate a runtime environment for local execution of cloud-compatible functions. In addition, two operating systems customized for edge devices with integrated Azure IoT connectivity, Sphere and Windows IoT Core, are available to component designers.

The recommended architecture and system design approach is to use the microservices and cloud-native serverless style of programming for the implementation of the functional modules that can be deployed and scale independently [15]. Recommended mode of communication is to use the REST-style services over HTTPS with JSON encoding.

Azure IoT Hub

The Azure IoT hub is a managed service that forms the core of Azure IoT platform [16]. Its main functions include:

- Device connectivity and data ingestion
- Data routing and service integration, rules engine
- Device twins (digital twins)
- Control plane: security, authentication, and management

Data Plane

Device connectivity modes include posting via HTTP and brokered messaging via AMQPP and MQTT in both native and web-socket variants. Communication between payload sources and consumers via brokers is in the publish-subscribe mode. This indirection decouples senders from receivers, provides durable storage of messages, and enables support of multiple receivers.

Received payloads are routed to their intended destination via defined rules or programmatically by the receiving modules. Messages can be filtered based on application properties, message system properties, message body, device twin tags, and device twin properties that are described later in this section.

Depending on the intended use, streaming data may be processed in flight and optionally stored in the warm storage, cold storage, or both. Routing is also a mechanism for integration and connecting with other Azure cloud services, such as analytics, event hubs, notifications, containers, and service bus queues and topics.

The Azure IoT hub maintains a device twin (digital twin) for each connected device. Device twin is a cloud representation of a device that stores its state information, metadata and configurations, and conditions. Technically, a device twin is a JSON document that contains reported and desired properties of the device, including its state. Device applications can detect discrepancies between the two states when they exist and initiate the required action to change the reported properties to their desired state. They can also receive explicit notifications from the back end to do so. The device twin also contains read-only device identity properties and tags that back-end services can read and write.

Device digital twins are designed so that they can be expanded with device object models that can form ontologies. They can also be used to create spatial graphs from multiple related devices and their reports that describe related aggregations and spaces of interest. Azure IoT also supports the concept of module identities and module twins that provide similar capabilities as device twins but with a finer granularity that extends to managing individual components of complex devices.

Control Plane: Security and Management

Security functions of an IoT hub include device authentication using per-device identities. Specific actions and requests from the authenticated endpoints and applications are further subject to authorization. Communications are encrypted at the transport layer using TLS or DTLS for TCP and UDP connections, respectively. Authentication is performed using X.509 certificates preferably from a trusted certificate authority (CA). Another supported method of identification operates through the use of security tokens that the hub can issue based on the unique device identity. Device credentials and identities for provisioned devices are kept in the identity store provided and managed by the IoT hub. To further enhance security and system integrity, devices are designed to not accept unsolicited network connections. Instead, they initiate all outbound connections. Moreover, devices should be configured to only establish connections with well-known sources with which they are peered, such as the Azure IoT hub.

The Azure identity store is primarily intended to manage security credentials via direct lookup. For device discovery and description, Microsoft recommends creation of a separate user-provided topology and entity store that may include device functional descriptions and models with related metadata. This store should be used by applications and services for constructing domain-specific representations of structural and functional relationships between endpoints and devices.

Device management includes functionality for different phases of a device lifecycle, including planning, provisioning, and retirement of individual devices and in bulk for groups. Device provisioning is a process of bringing a new device into the system that includes creation of its identity and security credentials. IoT hub can revoke credentials and disable devices and groups of devices that it deems compromised.

Azure Supporting Cloud Services

Azure provides a number of managed services designed for the general cloud use, some of which are applicable to implementation of IoT systems. Following roughly the system flow depicted in Fig. 8.2, they include Time-Series Insights (TSI) for stream processing and Cosmos DB for warm data storage.

TSI is an analytics, storage, and visualization service for the time-series data. It offers data aggregation and SQL-like filtering [17]. Time-series data are stored in memory and in solid-state storage for low-latency access for queries over potentially large data sets. It provides visualization in the form of graphs, tabular views, and heat maps that may be embedded into custom applications.

CosmosDB is a large-scale globally distributed "schema agnostic" database that supports multiple data models [18]. It may be used as a NoSQL-style document database. Azure SQL DB and Apache Casandra are also available for uses where their characteristics are a better fit.

Cold storage flows are usually directed to the Azure Blob storage or to the Data Lake service, depending on the specific requirements and preferences.

Business integration is usually directed to the Azure Logic that has connections and interfaces to the business intelligence systems, such as ERP and CRM.

User management, including role definition and separation, is commonly performed by the Azure Active Directory.

Azure Monitor is the tool for user-level monitoring of cloud applications and services [19]. It operates by connecting to device telemetry and logs to provide visualization and insights into the operation of a system.

Azure Applications and Services

Back-end services of most interest to IoT uses include visualization, analytics, and machine learning. In addition to various service-specific visualizations, Azure Power BI is a general visualization tool for its cloud services.

The Azure Machine-Learning service provides the tools and frameworks for training, deployment, and management of machine-learning models [20]. Models can be built and trained using several popular tools and libraries including Scikit, Tensorflow, PyTorch, and MXNet. Pre-trained models are also available for use as an open-source resource on GitHub. Data can be supplied by user or ingested from the cloud storage, such as the Azure Blob. Automated machine learning includes templates for streamlined model generation and training for common supervised learning uses, including classification, regression, and forecasting. Completed models can be deployed for execution in standard containers or on the Azure-provided FPGAs.

Models can also be deployed at the edge using IoT Edge runtime support for the Azure Functions serverless mechanism. Locally obtained insights can be sent to the cloud for notification, data integration, and model refinement as appropriate.

Azure IoT Edge

Edge devices can connect to the cloud via an IoT hub assuming they meet security requirements and can handle the required protocols and procedures. For less capable and legacy devices, Microsoft provides the Azure IoT Edge that runs on customer premises and outside of the cloud [21]. It provides device connectivity, a subset of IoT functions, support for local execution of applications, and limited operation when offline. The IoT Edge supports hardware security features, such as trusted platform module (TPM) and hardware security module (HSM), if available.

The IoT Edge runtime provides upward cloud connectivity to its attached or networked devices and performs communications gateway functions and protocol and format conversions if necessary. It manages secure communication with the cloud but relies on the IoT hub for device authentication when they first join the system.

The IoT Edge performs connection multiplexing for underlying devices over a common single connection to the cloud. This allows it to throttle traffic when necessary for congestion control. It can also isolate downstream devices from the Internet which may be required for added security by some OT installations, such as the factory automation or electric grid.

If necessary, data format and protocol translations may be performed at IoT Edge to make them suitable for sending to the cloud. For example, a segment of OT equipment connected via OPC UA, or a wireless network using CoAP, may be bridged to the TCP protocol stack. An IoT Edge provides a toolkit for writing protocol translators.

The second major function of the IoT Edge is to provide and manage the runtime environment for edge modules that implement its functionality. These may include data translation, filtering, aggregation, or edge analytics. Edge modules can be installed locally or downloaded from the cloud and connected to the back-end services. IoT Edge also monitors, manages, and reports operational state and health of its modules.

When connectivity to the cloud is lost, the IoT Edge hub goes into the offline mode for disconnected operation. In that state, it buffers messages destined to go to the cloud and authenticates modules for operation and communication among child devices that would normally go through the IoT hub.

Azure IoT Operating Systems for Edge Devices

The Azure IoT platform includes two operating systems that may be used at the edge to streamline development and cloud connectivity, Windows IoT core, and Azure Sphere.

Windows IoT 10 Core is a scaled-down version of Windows OS targeted for embedded devices that uses tools and interfaces familiar to Windows developers [22]. It provides security updates and Azure cloud connectivity.

Azure Sphere is an operating system designed for microcontrollers that power smaller and constrained edge nodes [23]. It is a secure operating system for real-time processing that supports cloud connectivity and over-the-air software and security updates. It works in conjunction with the Azure Sphere Security Service for maintaining secure, certificate-based connections to the cloud and web. Sphere OS is based on the Linux kernel modified to reduce size and to eliminate some common security exposures, such as the password authentication and shell.

Azure IoT SaaS and PaaS

Microsoft Azure IoT offerings also include SaaS and PaaS variants that are called Azure IoT Central and Azure IoT solutions accelerators, respectively.

Azure IoT central is intended to simplify the development of IoT solutions that do not require much service customization [24]. It makes use of Azure IoT architecture and services, such as the IoT hub and the basic analytics, without exposing their intricacies directly to the users.

The IoT Central's dashboard and functions can be accessed from browsers on PCs and tablets. It can be used to define and manage connected devices. The system supports multiple (non-programmer) user roles, including builder, operator, and administrator. A builder defines types, models, and templates for connected devices, including their telemetry and commands. A system operator can provision and monitor devices, configure rules, view telemetry and status, issue commands, and customize basic analytics functions and formats of their reports.

For all of this to work, custom code needs to be written to connect user devices and sensors to applications and the Azure IoT services. Microsoft provides several development kits to aid in this process.

Azure IoT accelerators, positioned as a PaaS offering, are complete IoT reference solutions for common IoT scenarios with open-source code that users can customize and deploy [25]. They focus primarily on the application part and connection with Azure IoT services. At the time of this writing, supported scenarios include remote monitoring, connected factory, predictive maintenance, and device simulation.

Other IoT Cloud Platforms

There are quite a few IoT platforms provided by vendors and open-source communities [26, 27]. Vendor-specific IoT platforms and clouds are available from a number of companies including HPE, Dell, VMware, Cisco, Siemens, Bosch, and others. Some of them rely on AWS or Azure IoT to provide specific core functions and services. They are usually closed systems focused on supporting the hardware and software from a single vendor, thus limiting their appeal as a general system-building foundation. They are intended to complement and extend a vendor's IoT portfolio and are promoted as a way to simplify customer deployment of IoT solutions.

Some emerging IoT software platform offerings are positioned as general-purpose solutions that do not depend on proprietary hardware. They often come from specialized software vendors that may not have significant track records with other products and whose longevity and potential market share may be unclear. They are proprietary, which puts their customers at the risk of having to do custom development and integration with tools and skills that may not be transferable in order to collect data that may not be easy to migrate to another system if a change becomes necessary.

The open-source offerings are somewhat fragmented, and some offer only partial solutions that need to be integrated with other components to complete the platform. The area is still nascent with no established dominant player.

In the remainder of this section, we briefly describe some elements of two other offerings from the cloud vendors, Google and IBM, and the General Electric Predix platform. Their designs basically follow the same conceptual framework of having the IoT core for secure data ingestion and device management, including registration and authentication. Input messages are typically brokered using MQTT and routed to the back-end cloud services. Message routing, aided by the variants of rules engine, channel the input data and connect them to a vendor's cloud services and user applications for processing.

Google IoT

One of the attractive features of the Google IoT platform [28] is its integration with the TensorFlow, a very popular framework for building machine-learning models. TensorFlow is a managed service on the Google Cloud, and it is also available as open-source code. Google has even developed a custom-designed chip, called Cloud TPU (TensorFlow Processing Unit), to accelerate model training and execution. TPU is an application-specific integrated circuit (ASIC) custom designed to speed up the execution of neural networks, both at the training and inferencing stages.

In general, ASICs have limited usability for the specific target applications but when designed properly can perform that task very well. TPU analysis conducted on the production system, used by more than 1 billion Google users worldwide, reports 30× to 80× performance improvements over CPUs and GPUs manufactured by comparable process technology [29]. In addition to the specialized hardware, those results benefited from some algorithm tweaks, such as using the 8-bit integer arithmetic to match the TPU architecture. Google has also made a scaled-down version of the chip, called edge TPU, available for integration into smart things and gateways to accelerate inferencing at the edge [30]. It is supported by the TensorFlow lite runtime.

IBM Watson

IBM provides public, private, and hybrid versions of its cloud with typical services, many of which leverage its Watson system for natural language processing, visual recognition, and machine learning. The IBM Watson became publicly known by winning a game of jeopardy in competition with human opponents. IBM IoT platform contains a functional equivalent of an IoT core which can funnel IoT data to the Watson AI system [31].

General Electric Predix

The General Electric Company is renowned for manufacturing the industrial equipment and control systems. It has made a considerable bet on introducing software to monitor and optimize such systems while in operation. The basic architecture of the GE Predix industrial IoT platform [32] is depicted in Fig. 8.3.

As indicated, the key core components of the platform are similar to those that we have encountered thus far. In addition to the customary sensor and edge data ingestion, marked as OT systems, it also explicitly calls out the asset data input since monitoring and managing of assets are one of the major concerns in industrial automation. The enterprise data and external data input boxes indicate that the system is

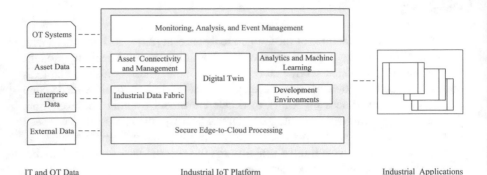

Fig. 8.3 GE Predix IoT platform

intended to integrate with other enterprise management systems for a holistic view and coordinated operation of the production and management systems, conceivably including the supply chain.

Due to some reorientation and repositioning of the corporate focus, the availability of the Predix platform outside of GE is somewhat unclear.

Open-Source IoT Platforms

Open-source IoT platforms have the similar conceptual design to what we have already described, with various amounts of their own or preferred specific components, such as protocols or database. Many of them can be installed on premises or in private and hybrid cloud modes. The main problem is that the space is rather fluid and experiencing growing pains, currently with no major players or enterprise-grade backers to instill the adopter confidence.

References

1. Lamarre, E. and B. May (2017) *Making sense of internet of things platforms* McKinsey [Online] Available at: https://www.mckinsey.com/business-functions/mckinsey-digital/our-insights/making-sense-of-internet-of-things-platforms (Accessed Dec 15, 2019)
2. AWS IoT Developer Guide (2019) [Online] Available at: https://docs.aws.amazon.com/iot/latest/developerguide/iot-dg.pdf (Accessed Dec 15, 2019)
3. AWS Simple Storage Service Developer Guide (2019) [Online] Available at: https://docs.aws.amazon.com/AmazonS3/latest/dev/s3-dg.pdf (Accessed Dec 15, 2019)
4. AWS DynamoDB Developer Guide (2019) [Online] Available at: https://docs.aws.amazon.com/amazondynamodb/latest/developerguide/dynamodb-dg.pdf (Accessed Dec 15, 2019)
5. AWS IoT Kinesis Data Streams Developer Guide (2019) [Online] Available at: https://docs.aws.amazon.com/streams/latest/dev/kinesis-dg.pdf (Accessed Dec 15, 2019)
6. AWS Simple Notification Service Developer Guide (2019) [Online] Available at: https://docs.aws.amazon.com/sns/latest/dg/sns-dg.pdf (Accessed Dec 15, 2019)

7. AWS IoT Analytics User Guide (2019) [Online] Available at: https://docs.aws.amazon.com/iotanalytics/latest/userguide/analytics-ug.pdf (Accessed Oct 15, 2019)
8. Amazon FreeRTOS User Guide (2019) [Online] Available at: https://docs.aws.amazon.com/freertos/latest/userguide/freertos-ug.pdf (Accessed Dec 15, 2019)
9. AWS IoT Greengrass Developer Guide (2019) [Online] Available at: https://docs.aws.amazon.com/greengrass/latest/developerguide/gg-dg.pdf (Accessed Dec 15, 2019)
10. AWS IoT Things Graph User Guide (2019) [Online] Available at: https://docs.aws.amazon.com/thingsgraph/latest/ug/amazon-things-graph.pdf (Accessed Dec 15, 2019)
11. AWS IoT SiteWise User Guide (2019) [Online] Available at: https://docs.aws.amazon.com/iot-sitewise/latest/userguide/sitewise-guide.pdf (Accessed Oct 15, 2019)
12. AWS IoT Events Developer Guide (2019) [Online] Available at: https://docs.aws.amazon.com/iotevents/latest/developerguide/iotevents-dg.pdf (Accessed Dec 15, 2019)
13. Azure IoT [Online] Available at: https://azure.microsoft.com/en-us/overview/iot/ (Accessed Dec 15, 2019)
14. Azure IoT fundamentals [Online] Available at: https://docs.microsoft.com/en-us/azure/iot-fundamentals/ (Accessed Oct 15, 2019)
15. Azure IoT reference architecture (2018) [Online] Available at: http://download.microsoft.com/download/A/4/D/A4DAD253-BC21-41D3-B9D9-87D2AE6F0719/Microsoft_Azure_IoT_Reference_Architecture.pdf (Accessed Dec 15, 2019)
16. Azure IoT Hub [Online] Available at: https://azure.microsoft.com/en-us/services/iot-hub/ (Accessed Dec 15, 2019)
17. Azure Time Series Insights [Online] Available at: https://azure.microsoft.com/en-us/services/time-series-insights/ (Accessed Dec 15, 2019)
18. Azure CosmosDB [Online] Available at: https://docs.microsoft.com/en-us/azure/cosmos-db/introduction (Accessed Dec 15, 2019)
19. Azure Monitor [Online] Available at: https://azure.microsoft.com/en-us/services/monitor/ (Accessed Dec 15, 2019)
20. Azure Machine Learning [Online] Available at: https://azure.microsoft.com/en-us/services/machine-learning/ (Accessed Dec 15, 2019)
21. Azure IoT Edge [Online] Available at: https://azure.microsoft.com/en-us/services/iot-edge/ (Accessed Dec 15, 2019)
22. Windows IoT Core [Online] Available at: https://docs.microsoft.com/en-us/windows/iot-core/windows-iot-core (Accessed Dec 15, 2019)
23. Azure Sphere [Online] Available at: https://azure.microsoft.com/en-us/services/azure-sphere/ (Accessed Dec 15, 2019)
24. Azure IoT Central [Online] Available at: https://docs.microsoft.com/en-us/azure/iot-central/ (Accessed Dec 15, 2019)
25. Azure IoT Solution Accelerators [Online] Available at: https://azure.microsoft.com/en-us/features/iot-accelerators/ (Accessed Dec 15, 2019)
26. Goodness, E. et al (2018) Magic quadrant for industrial IoT platforms [Online] Available at: https://www.gartner.com/doc/3874883 (Accessed Apr 15, 2019)
27. Gartner Industrial IoT Platforms Market (2019) [Online] Available at: https://www.gartner.com/reviews/market/industrial-iot-platforms/ (Accessed Dec 15, 2019
28. Google IoT cloud [Online] Available at: https://cloud.google.com/solutions/iot/ (Accessed Dec 15, 2019)
29. Jouppi, N. P. et al (2018) 'A domain-specific architecture for deep neural networks' Communications of the ACM 61(9), p 50–59.
30. Google edge TPU [Online] Available at: https://cloud.google.com/edge-tpu/ (Accessed Oct 15, 2019)
31. IBM IoT cloud [Online] Available at: https://www.ibm.com/cloud/internet-of-things (Accessed Dec 15, 2019)
32. GE Predix platform [Online] Available at: https://www.ge.com/digital/iiot-platform (Accessed Dec 15, 2019)

Chapter 9
Putting It All Together

This chapter summarizes key phases and considerations in designing and implementing an IoT system. Most of the details of component design considerations are covered in the respective chapters on edge, communications, cloud, security, and data models. The chapter provides a more holistic overview of the overall IoT system design process. The first few sections summarize the process and the design considerations, followed by a detailed case study of a complete project.

IoT Project Design and Implementation

Major phases in an IoT project life cycle may be roughly categorized as (1) definition of system purpose and objective, (2) design of the system architecture, (3) implementation, and (4) deployment and operation.

The process of designing and implementing an IoT system typically includes the following major steps and iterations:

- Definition of the objectives and purpose of the project
- Conceptual design and system architecture
- Implementation and infrastructure design
- Piloting and validation
- Deployment – installation, provisioning, activation
- Operation – monitoring and management, security enforcement
- Evolution – modifications and addition of functions based on the insights and operational experience

The design and implementation process of complex IoT systems is highly iterative, and many decisions in prior steps may need to be revised in the light of additional insights gained at subsequent stages. This extends all the way to running the pilot installations and analyzing the operational experiences from them. The resulting outcome may entail dropping or modification of features and changes of density

© Springer Nature Switzerland AG 2020
M. Milenkovic, *Internet of Things: Concepts and System Design*,
https://doi.org/10.1007/978-3-030-41346-0_9

as well as selection of types of sensors and components based on the cost-benefit analysis of their feasibility, complexity, and performance. The same evaluation may continue with the system in full operation. One of the lessons of the Internet applications is continuous monitoring of performance and effectiveness of the system in order to implement improvements that can be deployed in the field through software modifications. For the production level IoT systems, changes need to be thoroughly tested in the staging area or, where possible, operated for a while in the passive mode on the real system to gain confidence before their full activation.

Types of IoT Projects

There are different types of IoT projects that differ significantly in their scope and durations. It is important to clearly articulate the type and purpose of the project early on. This provides guidance on what needs to be done and on determining when the goal is reached and the project should end. In certain types of projects, depending on the stated objectives, some of the major steps of IoT system design itemized earlier can be simplified or omitted. Common types of IoT projects include:

- Learning exploration
- Proof of concept (POC) – test and validation of an idea
- Pilot – small-scale system deployment, validation of concept and design
- Production systems

A basic learning exploration of IoT systems usually involves connecting one or more sensors of interest to a cloud service for visualization and some basic processing. This may be accomplished fairly quickly and inexpensively by acquiring an edge prototype board or a maker board that has interfaces to the sensors of interest and that is supported by the vendor of the selected IoT platform. Such projects usually require little if any custom coding for the basic connectivity and integration with sample cloud services. Additional coding may be necessary for specialized functions if required by the project. Several commercial vendors provide discounted or free IoT platform services for such explorations.

Proof of concept (POC) is generally a focused IoT project whose purpose is to test a specific design approach in terms of feasibility or effectiveness in realizing an assumed benefit. Sometimes the term POC is used to refer to the creation of a demonstration vehicle to refine and promote what might be an already established idea to the management and potentially to customers for funding and sales purposes. In either case, POC projects tend to have limited scope and duration. Such projects are often of the throw-away nature in the sense that they are not intended to be built upon and extended for production uses later. Consequently, their design may be simplified by relaxing or eliminating requirements on certain aspects of the system, such as security, scalability, reusability, robustness, and – to a reasonable extent – stability.

The term pilot project usually refers to smaller test versions of systems that are intended for production use. Their purpose is to validate the design and effectiveness of a system prior to being released to production. Pilot projects may be implemented in test and staging settings that closely mimic the actual production environment. Alternatively, they may be installed and run in the production environment as a passive overlay that does not disturb or alter its operation.

In most cases, the objective of a pilot is to graduate to a production use. If so, its preliminary design should include consideration of all major aspects of the resulting final system, with the pilot implementation providing a scaled-down skeletal initial version.

Larger IoT systems intended for production uses usually embody most of the steps defined in the previous section. An overview of how they can be put together is the subject of the remainder of this chapter.

IoT System Design

Designing an IoT system is an iterative process that generally starts with articulating design requirements, followed by a system architecture and ultimately its implementation. Some of its key aspects are described in the sections that follow.

Definition of System Objectives

The first phase in conceptualizing an IoT project should is to define its purpose, primary objectives and intended use. Basically, it means stating clearly what the system is intended to do and what are the expected results of its operation.

In general, an IoT system collects and processes real-world data with some target objective, such as obtaining quantified insights, providing operational advice, or implementing a closed-loop control through automated actuation driven by analytics and AI algorithms. It is sometimes prudent to progress towards the higher-level automation through stages and to select the final system functions based on the observations of its behavior in operation. This can mean starting with the visualization to obtain quantified insights and to identify the most promising areas for optimization, then deploying and fine-tuning analytics by operating it in an advisory mode, and finally moving to the closed-loop automation using the AI tools and ML models trained for the targeted improvements.

The articulation of a system's purpose largely determines and defines its functions and design, starting with what is to be measured and controlled, the type and volume of data to be collected, where and how to collect them, and the types of services necessary to process the data in order to produce the projected outcomes.

Conceptual Design and System Architecture

While they are conceptually separate stages, in practice IoT system design may be influenced to a significant extent by what appear to be implementation concerns, such as the characteristics and topology of the physical environment in which it is to operate. For example, infrastructure conditions and distances of the various data capture points tend to largely affect the type and connectivity of sensors that can be used and how constrained the edge nodes and things may have to be in terms of their power and networking options. Consequently, the design process may iterate between the conceptual outline and implementation restrictions until a reasonable balance is reached that enables it to conclude.

Data Collection and Sensor Placement

The nature of and variety of data to be collected determines the type and placement of sensors that need to be installed. Some sensors work as passive taps, and others – such as flow meters – may require intrusive installation. The cost of sensors and difficulty of their placement may have to be weighed against the importance of data that they produce. Power and connectivity availability, or lack thereof, also tend to influence the final choice.

Required sensor density and location may significantly influence the choice of type and capability of edge sensing points, ranging from the simple raw sensors that need to be connected to a gateway to the smart things and sensors strategically placed at points where fewer physical phenomena may need to be tracked, but their data should be processed and perhaps even acted on to some extent locally.

Connectivity

Placement of sensors is largely determined by the location of physical phenomena to be sensed. Their topological distribution may impose specific networking options. For example, physically dispersed sensors in agricultural uses or open-pit mines usually require some form of wireless networking. Lack of electric power at the point of sensing mandates the use of batteries. The combination of the two may dictate the use of constrained networks and devices at least in the outer segments of such systems. Distance between nodes and communication gateways such as edge routers requires choice of networking technologies and topology that meet the range and bandwidth requirements – such as a routed mesh, star, or combinations of those. Another important decision in wireless networking is the use of licensed or unlicensed spectrum. Assuming reachability of telco cell towers with the appropriate support for IoT communication, such as LTE-M on NB-IoT at the planned location, this is a tradeoff between the managed guaranteed bandwidth and the recurring cost of service vs. unlicensed networks that may be installed and operated without additional charges. Availability of wireless edge routers of gateways, such as LoRa, is another factor to consider. Testing for interference and penetration if there are physical barriers at the planned deployment locations can provide useful insights in making the final choice.

Different types of wireless network radios and interfaces have varying power requirements and costs that impact edge node design and can be a factor to consider, especially if the volume of edge nodes is expected to be high.

The choice of the networking protocol is affected by what the wireless network may support and the networking layer at which it is intended to operate, such as the link or the network layer. In IoT systems, connectivity at some point usually converts to the full TCP/IP stack that terminates with the application layer possibly with an added IoT framework. One of the important network design decisions is which segments, and consequently nodes attached to them, will support what level of networking. As discussed in chapter "Communications", when compliant networks and protocols are used – such as CoAP over 6LowPAN – Internet-compatible payload may be encapsulated at the edge for transport and processing by different segments and layers. Alternatively, constrained wireless edge networks using any protocol may be terminated at the edge gateway and their messages converted and relayed to the rest of the system with a different protocol on the fully capable TCP/IP segments of the network.

Wired networks and segments generally have nodes with richer power and computational resources that enables them to participate in the networking and protocol designs of choice. In some applications, such as industrial control, edge devices and sensors may use variants of legacy protocols, such as Modbus or BACnet, that may need to be translated and converted to standard Internet protocols and chosen data formats.

Data and Control Flows: Function Placement

An important aspect of the conceptual design of an IoT system is the definition of data-processing functions that it needs to perform in order to fulfill its stated purpose. It needs to acquire measurements of the physical world and transform them into insights and ultimately actions. Data-handling functions include the usual data acquisition, filtering, formatting, storage, and aggregation for use by the services such as visualization and analytics. Depending on the location of the target services in system hierarchy, intermediate helper functions may have to be included, such as message queuing and routing, short-term storage for streaming data, event processing and notifications, etc. Data-processing services may include visualization, various forms of descriptive and prescriptive analytics, and domain-specific applications.

In general, determination of the functional placement of data and command flows in an IoT system involves complex interactions and tradeoffs that need to be made to achieve a balanced result that meets most of the requirements. It is usually an iterative process of modifications and adjustments to various parts of the system and design assumptions, including requirements.

The design of data flows deals with the requirements of getting the data to processing services that may be placed at different points on the route from capture to the final aggregation point. Security, networking, bandwidth availability and cost, latency requirements, frequency of data sampling, and availability of power and places for installation of gateways or fog nodes are factors that play a role in the decision on the functional placement and data flows. If actuation is to be supported, control flows need to be added to the mix.

For scalability and flexibility, it is preferable to use a modular design with a pipeline of simpler functions dedicated to the specific tasks. In general, it is a good practice to follow the cloud-style programming with the deliberate use of microservices and functions with serverless-style invocations that can be dynamically placed along the data route and executed where it is most effective.

Certain requirements and restrictions may impose some initial decisions that anchor parts of the solution. For instance, if high-frequency sampling of some inputs is required but only the resulting events when detected are of interest to the rest of the system, the most cost-effective approach may be to provide the processing capability at the edge and reduce bandwidth, latency, and storage requirements by processing such signals near the source. If local control loops are to be implemented for autonomy or speed of operation, they need to be implemented at the edge. If analytics and machine learning are to be applied to large aggregations of data, those functions should be placed in the cloud.

In actual practice, design decisions may also be influenced by nontechnical considerations, such as business relations, resulting in particular partner and vendor preferences. These may limit some component choices to the vendor-specific offerings.

Distributed Data Storage

If data are to be stored at intermediate points, storage distribution and placement of the related functions need to be defined accordingly. In some systems, raw data sampled at high frequency may be used to detect events of interest to the rest of the system, but also stored locally for archival purposes such as detailed analysis of data and local event sequences that preceded events of the particular significance. For example, tripping of safety switches in transformer stations of power-distribution grids may need sequencing of events that preceded it with millisecond resolution to determine the root cause of the problem. In such systems, edge nodes need to at least temporarily store the ordering of system events with the required time resolution.

If storage is distributed across the edge and fog nodes and the cloud, replication and synchronization mechanisms need to be defined to ensure consistency of data. If data are dispersed so that high-resolution samples reside at the edge, formalized retrieval queries and internode APIs need to be designed to support application access to them at the applicable levels of system hierarchy, including the cloud.

Data and Metadata Formats: Naming

A very important part of the design process is to select data and metadata formats for transport and storage. With no dominant IoT standard established in this space, it is good practice to emulate the general structure of information models and metadata annotations described in chapter "IoT Data Models and Metadata". This allows for easier mapping of data formats to specific standards should one be adopted later and for cross-domain aggregations. There is less established guidance for handling of metadata. The design process should specify which initial types of metadata are to be collected, how they are going to be named for semantic parsing, and how they are going to be generated and stored. Both data and metadata formats should be explicitly defined and documented, especially if they are expected to be preserved and reused across applications, domains, and perhaps other system incarnations in the future.

Naming of addressable points at the edge is necessary to identify sources of data and targets of control. Unique identifiers and their derivatives are commonly used to identify, address, and authenticate data sources in messages and as keys for storing and retrieving archived data. IoT systems that incorporate IP addressable endpoints may have their names resolved to addresses via DNS or through similar mechanisms implemented in a local domain. Dedicated IoT directories may be used to facilitate discovery and searches of things and objects in a domain. In any case, name management system and its operation need to be outlined at the design stage.

Security

Design of system security begins with a threat analysis and risk assessment that influence design requirements and inform implementation decisions. It closely interacts with networking in terms of defining security trust zones. If lower security is used due to constraints or lack of physical security at edge nodes, proper barriers and buffers need to be placed at the points of transition to the more secure parts of the system.

System security design also needs to cover levels of security and authority of different types of system users and authentication mechanisms for users and nodes. When digital certificates are to be used, appropriate mechanisms for their issuing and distribution need to be identified and made available internally or from third parties. Depending on the system security requirements, edge nodes may need to incorporate hardware assists, such as the trusted boot, secure execution environments, and encryption engines with secure key storage. As indicated in chapter "Security and Management", security design and implementation have an interplay with networking and choice of protocols that provide the required and specified level of transport security. Thus, it should be a consideration in finalizing the protocol selection on the various segments of the network.

System Implementation

While the conceptual design deals mostly with what and why, implementation is primarily concerned with how to reduce those requirements to practice. As mentioned earlier, system design and implementation stages tend to progress in iterative fashion and are not often clearly separable into distinct phases. However, since they have somewhat different objectives, in this section we discuss predominantly system implementation considerations.

Implementation is primarily focused on the mapping of system architecture to the actual components that will instantiate it. This includes things and nodes at the edge and fog and in the cloud that can carry out the projected end-to-end functional distribution.

System architecture and requirements inform the design of components with appropriate computation, storage, and communication capability to carry out the assigned tasks. It leads to the specification of characteristics of hardware and software components to acquire or build.

Edge

In terms of basic functionality, IoT edge nodes need to acquire data, perform local processing, and relay raw and processed data to other parts of the system as specified by the functional placement and data-flow design. Local processing may include simple operations such periodic data reporting, averaging, or notifications when specified thresholds are exceeded. More complex functionality may include data storage and edge analytics. Data sent to other system components need to be formatted according to the system data design and information model, with metadata annotation as needed. The resulting payload is forwarded using the designed method, push or pull, via selected protocols and serialization method directly to the receiving points or via an intermediary through messaging or a publish-subscribe system.

Key elements of the implementation of edge nodes include hardware, runtime environment, and applications. They are briefly summarized in the following sections.

Edge Hardware

Edge things and nodes need to contain relevant sensors and actuators and hardware for processing and communication as well as software to fulfill their designed purpose. At the minimum, the hardware includes processor with the required security features or enhancements, memory, input/output interfaces for sensors, and the chosen mode of networking. The range of hardware options for IoT components is very large. Depending on the functional requirements, processors used in IoT nodes can range from the small 8-bit microcontrollers and custom SoC components, all the way to the high-end processors with multiple cores, GBs of memory, and hardware support for virtualization and security that are normally designed for PCs and servers. Sensors and actuators also vary from basic transducers to complex subsystems with dedicated functionality that can be accessed via specified interfaces and protocols. Edge node hardware design or selection should include the hardware support for the chosen level of security, such as the secure generation and storage of encryption materials and keys, secure boot, and optionally hardware accelerators for encryption and decryption operations. In systems where edge nodes may be tasked with performing front-end analytic functions, their hardware design may include hardware accelerators for inferencing such as custom ASICs or FPGAs.

Many commercial hardware implementations of IoT gateways are essentially hardened PCs designed to withstand various degrees of harsh operating conditions, such as the extended range of operating temperature and resistance to dust and vibrations. They tend to have common PC interfaces – such as Ethernet and Wi-Fi, USB, and Bluetooth – with options for cellular connectivity and some combination of RS-232, RS-485, or RS-422 serial ports that are commonly used in industrial automation, but no particular provisions for direct interfacing of sensors and actuators. One way around this problem is to use commercial industrial sensors that provide hardware and software interfaces to components like programmable logic controllers (PLCs) that can be interfaced to a PC-based gateway using some type of its RS-xxx ports. Industrial control system tends to use legacy protocols, such as Modbus or PROFIBUS, to communicate with sensors and actuators. In such systems, suitable protocol translators need to be acquired or written to communicate with the gateway applications that process the data locally or convert them into the formats designed for use in the rest of the IoT system.

Edge Applications and Runtime Environment

Software and applications can be obtained or modified from existing sources, such as IoT frameworks that may provide open-source or licensed applications. As described in the earlier chapters, such frameworks usually provide the tools for the common edge operations, such as data posting via messaging and event processing, and some popular communications and security protocols. However, any system or domain-specific functionality needs to be custom coded.

In addition to the data-plane functions, edge nodes need to host and execute agents for security and management that communicate with their respective control and operation systems and consoles. Their code may be provided by the vendor of the chosen commercial security, management, or IoT platform. If the chosen edge platform and runtime system are not supported by the vendor, or if the control plane is custom developed, its edge agents need to be developed for each type of the edge node in a way that conforms to the security procedures and protocols used by the rest of the system.

Based on the required functionality and application-development preferences and choice of languages, runtime implementation options may range from the bare-metal support to a combination of an operating system with the required language libraries. Other considerations for the operating-system selection include real-time support capability if needed, support for the chosen processor architecture, availability of drivers for the chosen interfaces including network, and support for the required protocol stack, such as the TCP/IP, and for the specified security features and transport-level security protocols. High-end gateways and nodes may require additional support for virtualization, containers, and security features. Operating system size and cost in terms of maintenance and licensing terms are additional considerations that can impact the final selection.

General-purpose operating systems, such as Linux and Windows, provide rich functionality and have a wide range of drivers, libraries, protocol, and language runtimes in addition to a large base of developers. On the negative side, they tend to have sizable processing and memory requirements that may exceed the capability of the constrained edge devices. Embedded versions of both OSs may be possibly better suited to existing IoT uses. In principle, they aim for streamlined selections of features, schedulers that are tuned to better fit real-time needs, and elimination of unnecessary code. This results in lower demands on system resources and improved security by reducing the size of attack surfaces by removing unsafe protocols and entry points, such as the FTP and the command shell. However, those versions of OSs tend to have more limited distributions with different cadence of updates and support and may still be too large for edge nodes below gateways.

Native real-time embedded operating systems, some of which are specifically targeted for IoT devices, are an alternative for implementing things and smart sensors. They generally support a narrower range of hardware, runtimes, and protocols, but tend to be available for microcontrollers and SoCs popular in the IoT world. Their selection is subject to considerations similar to those described earlier, with the added caveats about availability of suitable development and test environments and long-term viability of the vendor or vibrancy of the open-source project that support them.

If the specific types of sensors and actuators to be used are not supported by the chosen runtime environment and OS, it may be necessary to develop custom drivers for them in a particular implementation. Due to a great variety of those, as well as of interfacing methods and runtime environments in use in IoT systems, this is the case more often than in the established ecosystems with more standards and dominant players, such as the PC environment.

Fog

Depending on the design of a system node and function hierarchy, there may be a need or a preference for an intermediate layer between the edge and the cloud that is commonly referred to as the fog. In general, fog nodes are a level up in hierarchy and tend to be characterized by connectivity to several edge gateways and having few if any direct sensor connections. Functionally, fog nodes may provide additional services to the raw and derived endpoint data and events that flow through them, such as detection and management of cross-node events, data storage, edge analytics, and visualization. These functions are similar to those provided by cloud, but they operate on subsets of system data and on smaller aggregations defined by the scope of endpoints and edge gateways connected to a particular fog node. In terms of implementation, fog nodes tend to have higher-end processors with generous amount of storage and hardware support for security and virtualization. They often support containers which, together with virtualization, provide mechanisms and tooling for dynamic function placement and migration between the fog nodes and the cloud in response to system needs and load variations.

Cloud

Major components of the cloud portion of an IoT system implementation include data ingestion, real-time stream processing, storage, security and management, and applications and services.

Compared to the wide range of choices for implementation of edge components, cloud functionality tends to be implemented by fewer but more complex and elaborate subsystems. Implementation considerations for the first four are briefly described in the sections that follow. Applications and services tend to vary considerably depending on the functions of an IoT system and on the vertical domain to which it is applied. They are discussed in chapter "Cloud" in relation to the cloud.

Data Ingestion

Depending on the projected volume of data to be ingested, a cloud implementation may need to include provisions for scalability, such as installing multiple brokers for MQTT systems or Kafka-like systems for reliable messaging and publish-subscribe forms of data delivery. For systems that use data posting to web servers, in formats such as HTTP, load balancers may be used to increase scalability.

Real-Time Stream Processing

If real-time processing of data streams is required in the cloud for rapid handling of significant events or timely detection of anomalies, some variant of lambda architecture should be implemented to deliver data both to stream processing services and to the storage for aggregation and post-processing.

In systems that have a lot of complex events, it is quite common to have a formalized tool for implementation of event processing. Such tools can process elaborate combinations of events and derivatives emanating from different parts of the system and direct them to final destination, such as the visualization, operator notifications or inputs to analytics, and AI services. Another common form of stream processing in the cloud is anomaly detection that monitors a number of inputs from different parts of the system to identify complex patterns indicative of developing problems that may require special attention and handling. There are no widely adopted solutions for those uses, so implementations tend to rely on the tools developed for similar purposes in the cloud and enterprise settings, or on bespoke developments.

Data Storage

The implementation of data storage needs to meet design requirements for volume, throughput, and scalability. As discussed in chapter "Cloud", these considerations coupled with varying data formats resulting from the potential addition of new types of devices usually lead to the selection of database types suited for or specialized for IoT, such as NoSQL or real-time streaming variants. In systems where data are distributed over multiple locations or copies, additional considerations include data synchronization, consistency, and convergence to a common state. An IoT database implementation also needs to address the handling and management of metadata. Data-management policies, such as retention and compression of older archives in terms of space or time also need to be specified and applied as appropriate in system operation. A database implementation needs to be able to efficiently support the types of queries that are common for the applications that it needs to serve, such as visualization, analytics, and AI processing. In IoT systems, databases need to be able to support data queries by time in addition to whatever other modalities may be required.

Applications and Services

Services and applications that operate on aggregated IoT data, such as analytics and AI/ML processing, typically reside in the cloud. Their development, training, and tuning for the production use may be a complex process that involves multiple steps as outlined in chapter "Cloud". If integration and interoperation with other enterprise application is required, such as CRM and ERP, these are also implemented at this level.

Security and Management

System and security management functions are usually coordinated and directed from the cloud. Their central operations, databases, and control points including operator management consoles and dashboards tend to be implemented in the cloud, usually as a separately functioning control overlay.

They manage system component throughout their life cycles and are designed to maintain them in secure operational states. The process of initial system commissioning and addition of nodes needs to be specified and implemented accordingly. When new nodes are added, they need to be entered in system inventory and directories as appropriate, provisioned with authentication and security credentials, and activated for operation. A management system needs to be put in place to keep track of the node status and to facilitate distribution of software and security updates necessary to keep them in compliance with system policies. It may also detect and disconnect rogue or compromised nodes to maintain system integrity.

System Integration

What to buy and what to build are important considerations in the implementation of any complex IT and IoT system. It is a complex decision that needs to be based on a number of factors that include the cost and time needed to complete the project, the likelihood that it will be repeated elsewhere for use in other plants or for sale, strategic importance, potential trade secrets contained in the process to which the IoT system is applied, and others.

In general, use of commercially available components that meet the design requirements reduces development and implementation times by providing ready-made elements that are supported and often come with software and tools for application development and integration. There is a wide range of commercial offerings for IoT systems, ranging from sensors, things, and gateways all the way to complete IoT platforms that include gateway software, ingestion mechanisms, storage, and services such as real-time event processing and AI frameworks.

The drawbacks of buying include potentially higher costs of acquisition and the recurring licensing fees that may be based on volumes of data. Buying also implies being tied to vendor-specific tools, procedures, and data formats that may limit portability and reusability of data across different systems and vendors. Choice of system components may also be limited to those supported by the platform vendor. Another cost and time factor to consider is the learning curve that may be needed for the developers to master vendor-specific tools. Those skills may not be transferable to other environments.

In addition to the common IoT subsystems that may be bought or built, the implementation of a complex IoT system usually requires incorporation of domain-specific applications and integration of all the piece parts to produce a useful functioning system. Integration involves fitting all the pieces together and ensuring that the intended flows are functioning and performing as specified by the system design. This may entail configuration and customization and, more often than not, development of some custom integration code even when using IoT frameworks and platforms. Depending on the magnitude of the project and the availability of the resources and skills, integration tasks may be performed internally or entrusted to an external systems integrator.

Deployment

Depending on the size and location of an IoT system, the initial system commissioning and activation can be a fairly complex undertaking. It includes installation of the component hardware and software, including potentially numerous edge devices, sensors, and networks, as well as fog and cloud nodes and services. It is common to start by activating at least a skeletal version of the cloud portion of services and management and control functions, followed by the gradual activation and integration of edge nodes. This process is referred to as the system commissioning.

As with integration, depending on resource availability, complexity, and size of the project, some of these steps may be entrusted to a systems integrator. If so, it is wise to have the same integrator perform both functions in order to avoid finger pointing in fixing the issues that invariably arise when bringing complex systems online.

Emerging Technologies

Several emerging technologies that are not explicitly designed for IoT may be useful and should be considered for new designs, subject to project objectives and their suitability in terms of availability and fitness for the purpose. They include virtual reality, augmented reality, and blockchain.

Virtual reality (VR) basically creates synthetic cyber worlds with simulated experiences in which users can navigate and manipulate (virtual) objects. In the current practice, virtual reality client settings consist of head-mounted goggle-like displays with small screens or a controlled setting with multiple projections on large screens. They can provide video and auditory renditions of virtual environments and may include other sensations, such as haptic feedback (touch) and vibrations. While the primary business driver seems to be the entertainment, virtual reality is finding uses in training and evaluation of system design concepts. The most obvious use in IoT and industrial systems is for training the personnel for complex procedures that may be encountered in operations monitoring, manufacturing, and service repairs.

Augmented reality (AR) provides a hybrid user experience by superimposing virtual elements over the physical world views. In its simplest form, it can provide virtual annotations of the physical world, such as a textual description of physical objects that are in the user's field of view. It typically works by providing overlays over camera images in a display, such as the smart phone or a tablet. In IoT uses, AR may be used to annotate objects for recognition and numerical information or to provide training and manipulation guidance. For example, an IoT system can be augmented so that it displays readings obtained from sensors that the camera is pointed at. In training and service uses, it can provide the relevant information about the operation or service procedures for the particular pieces of equipment that are in the user's field of view.

Enabling of AR annotations in IoT systems requires some means for detecting what the user's camera and display are pointed at. This can be accomplished in one of the two principal ways – by using precise location or image recognition. The location approach requires mapping of the space where AR is to be used. This means recording of the precise location of objects of interest and their identity. In addition, the user's precise location needs to be tracked, as well as the orientation of their camera – typically by using the smart-phone or tablet sensors – in order to pinpoint their focus of interest and to supply the appropriate AR information. This may include object identification and description and its status or sensor reports from the system streaming and stored data repositories. Precise location outdoors is provided by GPS sensors on user's devices. Indoor location is more challenging and may require the use of strategically located beacons.

In the image recognition approach, an ML algorithm may be deployed to recognize what the camera is pointed at, by using a database of the relevant object images that should be previously collected and classified. Given the relatively limited number of IoT objects in a particular system, the model training may be simplified by using the labeled sets of reference images and knowing that the outcome should resolve to identifying one or more of the known objects.

Elements of the VR and AR technology may be combined to create what is referred to as the mixed reality, i.e., a hybrid fusion of the virtual and physical worlds that incorporates elements of both. One of its mundane uses is to prevent users who are physically moving while their sensory inputs are generated by the virtual world, from bumping into objects in the real world (that they would otherwise not "see" in their simulated views).

Blockchain is essentially a distributed ledger that is designed to be tamper proof. It was introduced as a mechanism for tracking transactions of the digital currency, bitcoin. Blockchain keeps a securely audited trail of transactions in a distributed system with no central authority or mediator of trust. In its original incarnation, it operates as a public, permission less, and anonymous system. This means that entities posting transactions to the ledger do not have to obtain permission from some authority to do so nor do they have to reveal their identity. The ledger is effectively an append-only log, whose entries include a hash of the current transactions and of the previous ones. Details of secure hashing and authentication are discussed in chapter "Security and Management". From the functional perspective, blockchain is a distributed secure log of transactions that is unforgeable and may be verified by the interested nodes. It is maintained by a network of nodes in a distributed and decentralized peer-to-peer fashion.

Much of the excitement around the blockchain is conflated with its sibling the bitcoin. It is often overshadowed by the oscillations and market speculations of the bitcoin value, expectations of liberation of financial transactions, and new market opportunities that that digital currencies may provide.

The general blockchain as described here may not be a good fit for IoT applications for several reasons, most notable of which include computational complexity and the incentive system. The latter is provided in the bitcoin world by the process of "mining" where interested nodes provide elements of heavy lifting to maintain and secure the ledger in exchange for payments in bitcoin discoveries or transaction fees. However, variants of the blockchain may be devised to meet the IoT requirements. They can do so by eliminating the requirement to support bitcoin transactions and by reducing the computational and storage burden on the potentially constrained IoT edge nodes.

As a technology, blockchain and variants of distributed ledgers may enable several new uses and features in IoT systems. They include auditable tracking of components (and items of value in general) through their traversal of the supply chain from the point of origin, e.g., a manufacturer, to the final distribution. This can be especially valuable for the highly sensitive products, such as medicines and food. It is also very useful for tracking of shipments, such as containers and their content, both in terms of location of the goods and their authentication. Variants of blockchain have also been proposed as a mechanism in IoT systems to ensure the provenance of the collected data, i.e., that they actually originate from the sources that they claim to be. Another potential use of blockchain is for machine-to-machine payments, including very small ones that may be required in IoT systems that want to engage in trading of data and services.

The list above represents some of the more obvious uses in IoT systems. As stated, these technologies are emerging, so many more new uses remain to be discovered and implemented.

A System Case Study: Personal Office Energy Monitor (POEM)

This section describes the design, implementation, and user evaluation of an IoT pilot project that focused on monitoring and management of user comfort and energy usage in office buildings. The objective is to illustrate some key phases and considerations described in the previous sections by describing a fairly complex actual IoT project. It includes motivation and outline of the problem statement, the resulting definition of data to be collected, complete design of the implemented system, and the experimental and finding phase with analysis of effectiveness and value to users. The section closes with a brief discussion of project changes resulting in modification or omission of functions that were envisioned but not implemented for various reasons.

Motivation and Objectives

The project goals were defined by the Net-Positive Energy Consortium formed to explore the impact and synergies of technologies that could be deployed in constructing and operating buildings that are energy net positive, i.e., produce more energy than they consume. This is a challenging and important task since buildings consume approximately 40% of the total energy used in the USA [1]. The international consortium, now dissolved after achieving its stated objectives, included eight major global corporations engaged in all phases of the design, construction, and operation of smart buildings. Member companies had product focus and expertise in building construction, building-management systems (two), office furniture and space design, cafeteria management, solar division of an energy production company, computer equipment, and printer manufacturing.

One of the key conclusions of the member companies in the problem analysis phase was that the operation of energy-positive buildings will require awareness and active and willing participation of its occupants. Some internal studies projected that informed modifications of user behavior could save 12% of the total energy consumption of the building. It was also observed that in the current practice there is no direct connection between users and the systems that manage the building. As one BMS vendor participant observed, "BMS is supposed to manage the comfort and safety of building's occupants, but it has no idea where they are, what they are doing, and what their needs and preferences are."

This is a major missed opportunity, since maintaining the comfort of building occupants is the primary target of a building management system. Moreover, the users – by virtue of their actions and behaviors – are the primary driver of building's energy consumption.

The primary design focus of the project was chosen to provide the personalized information to users on their ambient conditions and create a mechanism for communicating their comfort preferences to the management system. A secondary objective was to measure and assess the aggregate and personal energy consumed by the IT equipment since no reliable data on that were available to the consortium to guide the smart building design and consumption projections. The specific initial focus was on the user's personal computer and printer usage. The project was named POEM for Personal Office Energy Monitor. Its focus was on monitoring and managing the demand side, with informed user involvement to minimize consumption, while maintaining user satisfaction, comfort, and safety. The supply side, i.e., energy production to achieve the net-positive energy goal, was part of another project not described here.

POEM Design Approach

Based on the stated objectives and some supporting research [2–7], the design approach was to implement the pervasive personalized sensing of the user ambient and IT energy usage. An early original slide of the conceptual design of the system to address these objectives is depicted in Fig. 9.1.

The plan was to collect electricity usage data by installing smart (Internet-enabled) plug load meters in the user's workplace. A somewhat uncommon choice was to attach the ambient sensors for the light temperature and humidity to user's PCs. This is convenient because it provides power, connectivity, and added

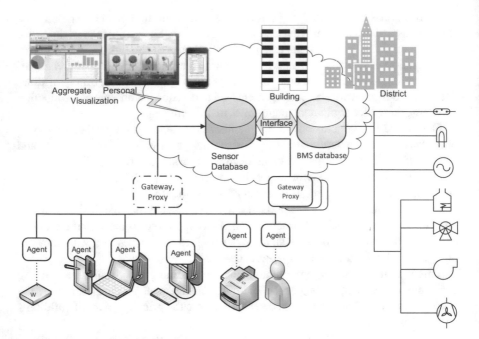

Fig. 9.1 POEM conceptual design

computational capability from the PC that the sensor is attached to. Moreover, users spend a lot of time near their PCs, thus making them a good place to measure personal ambient conditions. On the pragmatic side, one of the companies involved in the project was a computer company with interest in testing the use of sensors on PCs in an office environment.

The plan was to collect sensor data in a sensor database that would interface to the BMS system to report energy consumption and user preferences for ambient settings. The sensor database was also intended to provide data for the personalized visualizations of measurements of interest to individual users and the aggregate reports to system managers.

POEM User Interface Design

In order to define the system functional requirements, project designers started with outlining functions that need to be provided to users. Much of the attention was placed on the user interface design of POEM, both in terms of functionality and appearance. This took several months and intentionally preceded completion of the system architecture in order to make sure that it could provide all the desired functions. The user interface design went through many iterations, starting with whiteboard functional sketches, until it finally settled on the representation depicted in Fig. 9.2.

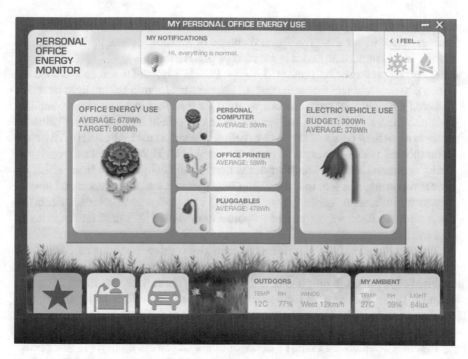

Fig. 9.2 POEM user interface, main (home) screen

The POEM UI screen contains fields for user energy consumption feedback, ambient condition dashboard, personal comfort feedback, and notifications. Energy data are represented both in quantitative terms and as intuitive visual indication.

Research [8] and our preliminary studies [9] indicated that users prefer to have an intuitive visual indication of how they are doing, as opposed to a numeric-only indication. The latter imposes the cognitive load of knowing what the good values are supposed to be, which is not always intuitive. For example, for the PC power consumption, it was not clear what the magnitude of the impact of user behavior and usage modality is – such as proactively initiating a sleep state when not in use – versus the baseline consumption dictated by the specific configuration and Energy Star rating of their unit [10]. Thus, a user dutifully following energy-saving recommendations could be consuming more energy in comparison with another one who is not if their respective PCs differ significantly in terms of their inherent energy efficiency. In order to encourage good behaviors, the POEM personalized indicators and feedback had to be adjusted accordingly, without burdening the users with the raw data that they would have to scale as appropriate for their particular PC.

To deal with these issues, we opted for using a visual metaphor to depict the basic "bad, neutral, good" ratings. In our design we opted to use the shape of a flower with a flourishing image depicting the good and a withering flower corresponding to what was in effect the bad state. We renamed that state to something meaning "could use improvement" in order to strive for the positive reinforcement and not to discourage users. We found that the use of a visual metaphor works well and resonates with users. The specific choice of a visual representation, such as a flower, is less important, and it is a matter of preference.

Other key information presented on the POEM's home UI screen included the user's personal ambient measurements of temperature, humidity, and light intensity, "where I am right now," and outdoor weather for the exact location of their building obtained from an Internet-based service. A separate field labeled "I Feel" allowed users to input their personal subjective feeling of comfort using the ASHRAE seven-point thermal scale (−3 very cold to +3 very hot) [11]. This information was aggregated, annotated with the identifier of the heating and cooling zone from which the user is reporting, and sent to the building manager as a summary report and optionally as a notification, depending on severity. The plan was to later interface this input with the BMS for automated response to mediated user preferences in zones that can be individually controlled.

The field labeled "My Notifications" provides a mechanism for individual and group notifications of users from system operators and building managers, such as ambient adjustments in progress or estimated times for service restoration in case of failures. It can also be used for bolstering energy-saving efforts and campaigns to achieve specific individual and group goals, as well as for implementing challenges and competitions.

Figure 9.2 depicts the POEM's initial or home UI screen. It allows for user selection of additional screens, such as the electric vehicle charging and management. Individual energy fields are also interactive (clickable) to show details of various aspects of energy consumption. One such screen for office energy totals and comparisons is depicted in Fig. 9.3.

It shows a user's daily energy consumption for the past week and the computed average that is shown on the main screen. Various colored dots show energy consumption and comparable averages of interest to the user – user's department, floor, and building. Our preliminary studies indicated that users can relate to data much better when provided with a point of reference, indicating how their consumption compares to others that may be relevant. As an intuitive example, knowing that you printed 320 pages last month is of some value, but knowing that the average for your peers in the department was 130 gives you a much more useful and perhaps actionable frame of reference. User anonymity and privacy in such reports can be preserved by showing the individual data only to the originating user, and only group ranges or averages for everything else. The horizontal gray line indicates the target objective that the user is aiming for.

The visual indication of user's progress, depicted by the shape of the flower, was computed based on the difference between the target and the actual consumption. Target values could be assigned by users or by the system, depending on the policies in effect. Our strong preference was for the user-settable targets with the management policy settings as a fallback for users who were confused by options or did not want to make a choice themselves.

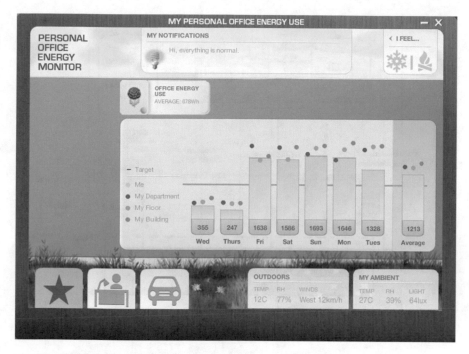

Fig. 9.3 POEM energy usage and comparison screen

System Architecture and Implementation

The focus on personalized user feedback and functionality drove many of the system design requirements. In addition to the typical IoT sensor data acquisition and processing, it was necessary to support attribution of measurements to individual points of origin, users, and groups of users determined by POEM categories. This was accomplished by the extensive use of metadata to annotate inputs and coordinated handling of data and metadata at the back end to maintain the required associations.

Other major aspects of the conceptual design included placement of sensors near users, on their PCs for ambient sensing, and on their power plugs for energy consumption. Edge nodes were collecting sensor data and providing the user interface, with most of the data processing and storage delegated to the back-end servers. Connectivity and security were basically provided by the network connections and protocol stacks of the host PCs, and the corporate IT mechanisms and policies provided by the enterprise networks on which they were installed. Data and metadata storage requirements included archiving of sensor data streams and of metadata needed to attribute and to report the measurements on both the individual and group levels.

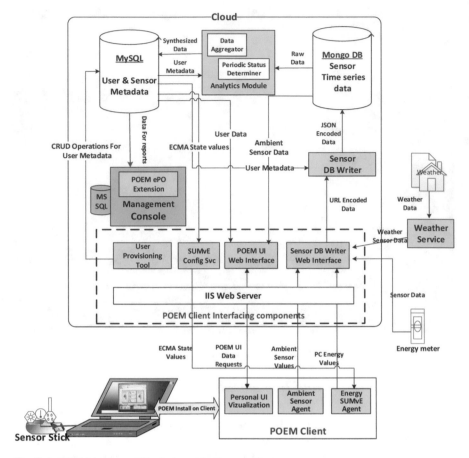

Fig. 9.4   POEM system architecture and implementation

The final system design as implemented is illustrated in Fig. 9.4. It indicates that the POEM system software consists primarily of the edge node software and the back-end server software.

Edge Nodes

POEM edge nodes are a user's PCs fitted with ambient sensors and stand-alone plug-level power meters. The original plan was to install power meters on the relevant office power plugs to measure the per-user energy consumption. During system development, a computational method was devised for determining PC energy usage based on the power-state occupancy. This eliminated the need for the physical energy meters, and they were installed only for the printers. Details of operation of the energy management sensor, called SUM vE, and its software implementation are presented in a later section.

The edge node software, labeled POEM client in Fig. 9.4, consists of the three functional modules:

- Personal user interface (UI) and visualization
- Ambient-monitor sensor agent
- Energy sensor and agent, SUM vE

The UI module is in charge of driving the user interface, including the display of relevant data such as depicted in Figs. 9.2 and 9.3, and supporting the interactions for user inputs. It obtains values to be displayed from the back-end server complex using specified queries.

An ambient-sensor agent communicates with the locally attached USB stick that contains the physical sensors for measuring temperature, humidity, and light intensity. The sensor board used for this purpose also includes an embedded microcontroller programmed to implement a subset of USB protocols to communicate with the generic Windows sensor drivers running on the client. The agent builds upon those and implements the interfaces and the logic to read values from the sensors of interest for the subsequent forwarding to the back end.

The energy sensor module contains a service that runs once every second to observe and record the energy state of the machine that it is running on. It uses consecutive aggregations of those states to compute the energy usage for the subsequent forwarding to the back end.

POEM Back End

The POEM back end is a collection of services that perform various functions, and a few interfacing modules for edge interactions, data ingestion, and miscellaneous services.

Some of the major back-end services depicted by Fig. 9.4 include:

- Sensor database
- User and metadata database
- Analytics module
- Management console

In addition to those, some simpler services provide helper functions. They include a sensor DB (database) writer that decouples sensor agents from a specific database implementation, a user provisioning management tool for adding system users, and a SUM vE configuration service described later.

System Flows

A POEM endpoint becomes operational when the client software is installed, the ambient sensor stick is inserted in the machine, and the user is provisioned, that is, a user ID is assigned and the corresponding static portion of the metadata are entered in databases using the provisioning tool described later. Ambient sensor data are sampled at regular intervals whose length is programmable. They are reported using another programmable interval, in our initial configuration set to 30 s. Energy sensor data are also reported every 30 s. Details of its operation are presented in a separate section on SUM vE.

When the user machine is active, sensor data are posted by the client software in regular intervals using the HTTP post to a dedicated URL that points to an IP and web server application port of the sensor DB writer-web interface in Fig. 9.4. This is an entry point to a web service that acts as the front-end of the sensor database writer helper service. Its function is to write the sensor data in the database and to decouple that operation from the sensor agent. This indirection provides the flexibility to change the database implementation at a later date without impacting sensor agents in the field.

Posted data are humanly readable strings in the rather generic HTML-compliant encoding of name, value pairs in the format that followed the spirit of some related early academic work [13]. A partial example of the POEM reporting format "on the wire" is illustrated below as the "name"-:"value" pairs in JSON-style format:

```
{"id":    "userID",    "tsutc":    "mm-dd-yy-h:mm:ss",    "light":50,
"tempC:"25, "rh":60, "indicator":"ambsensor"}
```

Names used in the name-value pairs were defined in the POEM vocabulary for consistency in referencing and parsing. The "id" field is used to designate the unique "userID" identifier associated with a PC as generated in the provisioning process. "tsutc" is the time stamp in Coordinated Universal Time (UTC) of the reported sample, and "light", "tempC", and the "rh" are names of the corresponding numeric values of the light, temperature in °C, and relative humidity in %, respectively. Comma is a separator and empty spaces are ignored by the data parser. UTC time was used for consistency across geographies, but the data were converted to local time zones for presentation to users and for analysis.

The example indicates that particular engineering units were implicitly assumed for the ambient sensor reports of the light, temperature, and humidity. In retrospect and with the benefit of learnings described in chapters "IoT Data Models and Metadata" and "IoT Data Standards and Industry Specifications", a cleaner design would have been to designate them explicitly via separate metadata fields.

As described, sensor readings are reported as a push operation, as opposed to a pull that would require a specific request from the server complex to each client. No special provisions were made for scalability of data ingestion, since the number of clients in our studies was relatively small – on the order of tens of users – and their aggregate data rates were well within the web server's ability to handle posts. The plan was to support higher aggregate data rates, when needed, by using a load balancer.

In POEM implementation, user ID is mappable and directly correlated with a stream ID which is used as a key for storing and retrieving data streams in the sensor database. Metadata annotations are used to label and to separate the individual data streams emanating from the particular sensors. All other (static) metadata for the user were stored in the relational database in a manner retrievable by the user ID, so they did not need to be reported with individual sensor updates.

In addition to energy and ambient reports, two other sources provided data to the sensor database – power meters and weather service. Commercial power meters had the capability to post data to a programmable IP address in programmable intervals. They were configured to report to a dedicated port in the sensor DB writer-web interface module. The program servicing that port parsed the data and forwarded the values of interest, such as the time-stamped current power draw and cumulative energy use, to the sensor database writer.

The weather service functional module was designed to query a public Internet service, such as the Weather Underground, for the outdoor weather conditions of the specific location where the user building is located. This information was obtained and posted to the database via the sensor DB writer-web interface on a regular basis with programmable intervals, typically 15 min. It was retrieved from there for display to the users and was available as an input to BMS if desired.

Data and Metadata Storage

Due to the extensive use of metadata and their largely different characteristics and update patterns from the sensor data, a decision was made to split the functionality between two different types of databases. Among other things, this would allow for the separate evolution of data and metadata implementations as understanding of the system improved by experiencing it in operation.

Per-user metadata included designations of user's department, floor, and building. Since this portion of user metadata is fairly static and it changes rarely if at all, a design decision was made to store those as tables in a relational database, with a unique user indicator as a key, and other membership fields defining the static schema representing memberships and associations of interest, such as a user's department and floor. In addition, sensor data were annotated with metadata to indicate the type of measurement being reported, such as the temperature, humidity, light, and energy. The primary user's office was included as their location in the static metadata since we could not find a suitable indoor location technology that worked in all pilot locations. The alternative of having the users input their location as they moved around with their laptops was deemed to be too imposing in terms of making the POEM application too chatty and potentially annoying.

Metadata storage, due to its mostly associative relations and relatively static nature once defined, was delegated to a traditional relational database. In the implementation described here, MySQL [14] was chosen because it met the requirements and had a license-free open-source version which was preferable for a pilot. Its potential transactional limitations on throughput were not of great concern given the volume and access patterns of metadata (predominantly reads) that were anticipated.

The sensor database implementation requirement was the ability to store streaming data without imposing a strict schema on the data, since we had a several different types of sensors – hardware, software, and human input – that we expected to grow by adding new types of sensors to be defined over time. For those reasons and others described in chapter "Cloud", a document NoSQL database was chosen, in this case Mongo DB [15]. The system used unique identifiers that served as keys in both databases so that the related entries could be directly correlated. For example, the relational metadata database used this key to identify the data streams and records in the sensor database related to a specific user.

From a practical point of view, a single database for sensor data and metadata may have been preferable, but we could not find any that would meet our requirements at the time when the project implementation begun.

Analytics

The analytics module operated periodically to compute group averages of interest. Every 15 min it computed individual usage trends based on the power consumption in the last cycle compared to the consumption for the corresponding part of the previous day or week average. The corresponding trending indicator is shown as a large colored dot (green) in the lower-right corner of the flower images in Fig. 9.2. Green color was used to indicate good trending, red for bad, and orange for marginal. Results were stored in the relational database and made available for display in individual POEM UI screens of the users when requested. Daily averages for all users were computed at midnight and made available to the UI as historical individual and group averages of interest to be displayed as illustrated in Fig. 9.3.

The analytics module was operated in 15 min and in daily intervals to compute aggregate data of interest and to store them in the database for user queries and archival purposes. It retrieved raw data from the time series in the sensor database and computed data for the relevant associations and groups, such as departments, floors, and total building data.

Individual user data are furnished to UI portions of individual active users. The POEM UI, when the application is active and displayed on the user's screen, requests data from the POEM UI web interface service. It in turn queries the two databases to obtain relevant data and comparisons and furnishes them to the UI. The reverse data flow, from the user to the database, is supported for users to provide feedback on their subjective feeling of comfort.

Management Console

The only major remaining system function depicted in Fig. 9.4 is the management-console interface. It was designed to provide the reports of interest to system managers, such as the IT and facilities (building manager). Reports included energy usage per user, department, floor, and building. Results of the pervasive ambient sensing were made available to the building manager in order to allow identification of rogue zones that were too hot or cold. Depending on their density, this information may or may not be captured by the BMS sensors, such as thermostats. Moreover, personal ambient sensors provided an independent corroboration of data. Significant discrepancies between the two systems, if encountered, can be used to identify malfunctions in either.

Implementation of the management console and visualization can be a fairly complex task. In order to save the development time and cost, we opted to use the configurable reporting feature of the existing management console used by the IT department for management of security and antivirus functions. It was relatively easy to interface to sensor and metadata database queries that in turn drove the visualization in terms of customary graphs and bar charts. It also eliminated the need for introduction of another dedicated console that system operators usually resent.

System Energy Sensor: SUM vE

This section describes some implementation details of a software agent designed to measure the PC energy consumption. It is included as an illustration of sensors that can quantify physical phenomena by inference from observing related indicators that are easier or cheaper to measure. Readers not interested in the details of SUM vE operation may omit this section without the loss of continuity.

The original design intent was to use the plug-load power meters to measure PC energy consumption. In the course of research and gathering of preliminary data, the design team discovered a mechanism that can be used to accurately estimate power consumption by software. The insight was that PCs have a small number of distinct power states in which their energy consumption is fairly constant. The idea came from the ECMA-383 standard [12] that defines a methodology for measuring the energy of personal computing products that is used by the Energy Star program [10]. It defines the following expression for estimating the total annual energy usage of a PC:

$$ \text{TEC}_{\text{estimate}} = (8760/1000) * \left[P_{\text{off}} * T_{\text{off}} + P_{\text{sleep}} * T_{\text{sleep}} + P_{\text{idle}} * T_{\text{idle}} + P_{\text{sidle}} * \left(T_{\text{sidle}} + T_{\text{work}} \right) \right] $$

where TEC is the total energy consumption estimate, P_i is the power consumption in a particular power state, and T_i is the fraction of the time that the system typically spends in that state. Power states of interest are "off," "sleep," "idle" when the system is on but the screen is off, "sidle" which stands for the short idle when the system is on and the screen is on, and "work" when the system is in the active mode, i.e., executing a workload, and the screen is on. The multiplier at the beginning comes from the number of hours in a non-leap year, 8760, divided by 1000 to express the energy in kWh.

Power state occupancy is detectable by software through a Windows system call. The energy spent by each power state is measurable or generally available from the manufacturers in their Energy Star specification. The POEM team modified an internally developed system utilization monitor (SUM) that collected PC usage statistics to detect and record power state occupancy. It ran every second and recorded the system state, such as idle or short idle. The energy monitoring agent accumulated the state occupancy over its reporting time, and it was then multiplied by the power consumption in each recorded state. That is practically a constant value that can be determined by using the external power meters to measure the power draw of the system in that particular state.

The SUM program tracked the length of occupancy of idle and short idle states. Sleep state duration was derived from the system event log which records the times when going to sleep and waking up or booting. Our own studies indicated that there is virtually no difference in the power drawn between the short idle and the working states. This allowed us to use a single-power measure for both and to avoid the problem of accounting for different workloads. A similar conclusion was reached by the ECMA and the Energy Star who also substituted the working state with short idle for similar reasons.

The SUM vE agent reports data in a similar "name":"value" format as the ambient sensors shown earlier, with user ID for correlation, time stamp, energy data, and a metadata indicator of the sensor type.

Figure 9.5 illustrates the results of validation of energy sensor accuracy performed on several machines in a typical office used by comparing the SUM vE computation to the actual consumption measured by an external plug-load meter to which the power cord was connected.

Some additional tweaks involved in completing the SUM vE (version Energy) implementation included adjustments for the external monitors and the laptop battery charging. External monitors are used by all desktops and some laptop users, typically docked in their office. Monitors are on standby in the long-idle and in the short-idle states, so the total systems power consumption may be computed by using their respective on/sleep/off power ratings or measures. For the monitor power tracking, their type designation needed to be included in the user information during system provisioning.

Since the measurement of interest was the PC energy consumed from the building's power distribution network, SUM vE did not report energy consumption for laptops while running on the battery power. However, it did measure the state of battery charge when the system is plugged in, to account for the significantly larger power draw of laptops when charging batteries. This is typically on the order of two to three times higher than the consumption when plugged in with the battery fully charged.

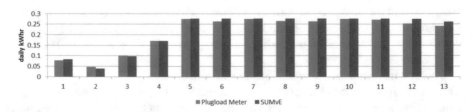

Fig. 9.5 Energy monitor accuracy, SUM vE compared with plug-load meter

The use of SUM vE for energy quantification and reporting simplified deployments of the system and saved the added cost of acquiring power meters.

System Deployment and Provisioning

A number of versions of the system were constructed and used for experimentation, development, and testing. The final version depicted in Fig. 9.4 was deployed in the two pilot studies with actual office users whose results are described in the subsequent section.

Deployments consisted of the edge nodes represented by the user's office PCs fitted with the POEM client and equipped with ambient sensors connected via a USB port. The server part was designed to run either on premises or in the cloud. Our initial implementation was on the servers dedicated for development and testing purposes. We used virtualized environments on local servers to facilitate porting to a commercial cloud that was completed later. Physical servers were initially located in the research laboratory and were later moved to a professional hosting facility for the benefits of professional IT management and Internet addressability that was necessary for implementing pilots on external corporate networks. A cloud version was later installed and operated using a commercial cloud provider.

Edge and end-user system provisioning process consisted of installing POEM hardware and clients on the participating machines and entering user information in the metadata database. We developed a user provision tool depicted in Fig. 9.4 to simplify the process. Per-user data included user's system ID, office location, department, floor, and building. Additional information included power-state draws for idle and short idle for the user's particular machine type. This was used to compute the energy consumption by the SUM vE energy sensor, as indicated by the query line labeled ECMA State Values in Fig. 9.4.

Pilot Deployments and User Study

The POEM system was piloted with the actual office users in two separate locations for approximately 3 months of continuous use. This was followed by a user study and extensive participant interviews performed by the human factors experts to assess the user experience and feedback on the select system features. The overall goal was to explore the ways of engaging users to actively participate and manage their behaviors in ways that might capture the projected 12% energy savings in operation of energy-efficient buildings with a view towards net-positive energy buildings.

More tangible specific objectives of the pilot deployments were to:

- Close the loop between building occupants and management systems
- Explore methods that might increase user engagement in building operation and energy usage through control of ambient conditions

- Explore the value of pervasive ambient condition sensing and interactive user feedback on personal comfort to building managers and management systems
- Quantify energy usage of IT equipment in office buildings for design and operation of net-positive energy buildings

Pilot Settings and Data Collection

The pilots were conducted for approximately 3 months with 27 users (regular office workers) in France during the spring and 46 in Japan during the heating season. Data were collected for the individuals, and aggregates and averages were compiled for the departments and floors. Personal data were provided to individual users via the POEM UI and to building and facilities managers via the management console.

As an illustration of some of the collected data, individual daily readings of temperature and humidity are provided in Figs. 9.6 and 9.7, respectively. Different days are represented by different colors. Mean laptop energy consumption by the department for one pilot is shown in Fig. 9.8.

Humidity is important as it, coupled with the temperature, impacts the perception of user comfort. Its indoor extremes can create a feeling of extremely damp or dry environment. Most of the readings in the two pilots were within the recommended range of 30–60%. This provided a quantified confirmation that systems were operating in the desired range.

Light measurements were collected for reference and potential future control uses.

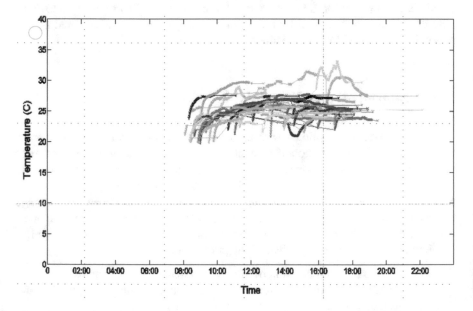

Fig. 9.6 Personal ambient temperature, daily readings for an individual user

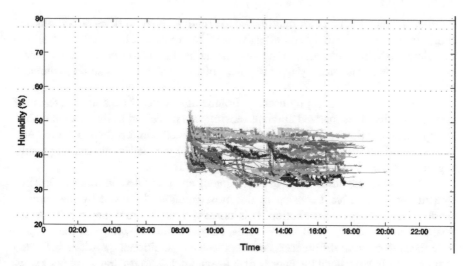

Fig. 9.7 Personal ambient humidity, daily readings for an individual user

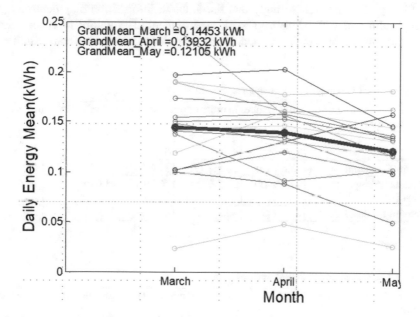

Fig. 9.8 Mean PC energy consumption, by department

Pilot User Study

Following completion of the pilots, extensive exit interviews were conducted to understand user experiences and their perceived value of system features. Some of the major findings were that users were very interested in having the personal ambient data and energy usage data, especially when coupled with guidance on how to manage them. Ambient data was regarded by users as very useful for organizational culture, and they felt that they contributed to creating better working conditions and might improve productivity and effectiveness. Many users expressed the desire to contribute to helping the environment by managing their use of energy in the workplace and were frustrated by the prior lack of metrics and tools to do something about it.

Table 9.1 shows ranking by users in France of key POEM features in terms of perceived value. Personalized ambient sensing was perceived as the most valuable. Energy consumption came in second and personalized comfort feedback the third. It was later discovered that this was probably due to the inadequate training resulting in users not knowing how to use this particular feature.

Comparable responses from users in Japan are presented in Table 9.2. They ranked personal ambient sensing as the most valuable, followed by the comfort feedback and energy consumption. It is unreasonable to draw major conclusions based on the relatively small sample sizes, but the intuitive hypothesis was to attribute these to cultural differences. Answers to some additional questions indicated that users in Japan valued the energy data as well but were frustrated by not knowing what to do about it and how to change them.

Pilot participants felt strongly that the ambient information may contribute to their health and well-being in a work environment. The ability to change their environment was also highly valued. From the point of view of the consortium that

Table 9.1 User rankings of POEM features, France

Please rank the following POEM features in order of priority: (1 = most valuable/used; 3 = least valuable/used)

	1	2	3	Rating Average	Response Count
Energy Consumption	31.6% (6)	**36.8% (7)**	31.6% (6)	2.00	19
Humidity, Temperature, Light Data	**47.4% (9)**	26.3% (5)	26.3% (5)	1.79	19
Comfort Feedback Tool	21.1% (4)	36.8% (7)	**42.1% (8)**	2.21	19

Table 9.2 User ranking of POEM features, Japan

Please rank the following POEM features in order of priority: (1 = most valuable/used; 3 = least valuable/used)

	1	2	3	Rating Average	Response Count
Energy Consumption	16.1% (5)	19.4% (6)	**64.5% (20)**	2.48	31
Humidity, Temperature, Light Data	**48.4% (15)**	38.7% (12)	12.9% (4)	1.65	31
Comfort Feedback Tool	35.5% (11)	**41.9% (13)**	22.6% (7)	1.87	31

initiated the study, these are encouraging results. For the construction companies and office space lenders, this translates to the increased value and competitiveness of user-friendly spaces. It also allows companies to demonstrate by actions that they care for the comfort and well-being of their employees in the workspace – always a good thing to do and a potential competitive advantage.

Aggregate data on energy usage provided a quantification input as a reference to building designers in the consortium. Data reports on the ambient conditions were highly valued by building managers. They provided an independent confirmation of the BMS-reported values and provided much more pervasive localized information that uncovered some problematic zones in buildings that tended to get too hot or too cold at times, due to the window orientation and sun exposure. After the initial hesitance to make the building data available to users, building managers found it extremely valuable to have the personal comfort feedback information and "being able to manage conditions at the level of a human being, as opposed to the impersonal facility management."

On the compliance side, the system helped with compiling reports on employee health conditions that are required in France and energy use and annual reporting on energy uses and savings that are required in Japan.

The final encouragement came at the end of the pilot studies, when many users were lobbying to have the system remain operational and did not want to remove it from their machines.

Implementation Notes

The previous sections described the POEM project as a complete example of a sizable IoT installation designed to collect targeted data, carry out planned experiments, validate some hypotheses, and provide quantified insights.

In the creation of POEM, the conceptual design and implementation phases were clearly separated, admittedly perhaps less attributable to virtue than to serendipity. Namely, the team first designed the concept and created an interactive mockup of the UI using a software tool to sell the idea and the concept for funding. Following the approval, it took a team of researchers and developers a better part of a year to actually reduce the concept to practice and to implement the system [16].

Some of the features that were initially anticipated turned out not to be workable or worthwhile and were excluded from the implementation. They included indoor location, plug-load disambiguation, electric vehicle charging, and print tracking per user as additional components of the more comprehensive quantification of a user's impact on the environment.

Indoor location was intended to localize tracking of the ambient conditions as users are moving around the office space. We found no reliable commercial or stable research solution that worked in all pilot settings. The fallback position was to install the ambient sensors near the user's desks and in the docking stations and to not have them report the data when their host machine moved.

Plug-load disambiguation works by analyzing the power signatures of measurements produced by the power meters to determine which particular devices on their network are turning on or off. This was meant to indicate when personal office lamps, printers, or heaters were on or off. The research available at the time did not provide a reliable way to accomplish this, so the decision was made to omit the feature.

Tracking of the electric vehicle charging was dropped since the buildings in which the pilots were conducted had either very few charging stations or no parking lots. Consequently, the effort required to provide this feature was deemed to be not worthwhile in the view of the miniscule potential benefit to the project.

Only one of the office environments in which the pilots were located had a printing solution that allowed for accounting of individually attributable printer usage. Our initial studies indicated that users were interested in this data, especially with the POEM feature of providing comparisons with the user-relevant averages as was done in the case of energy usage. However, interfacing to this capability required access to the user directory and authentication mechanism, and we did not obtain the permission from the management and the IT department to do so.

Collection of the light data was inspired by a study that indicated that user-directed lighting based on their needs and availability of natural light can improve user satisfaction and save energy [17]. However, we did not have a convenient setting to replicate this capability in the facilities where the pilots were conducted, so the light control was deferred to future versions.

These specific details were project specific and not generally relevant to IoT systems. Their descriptions were included to illustrate that changes based on additional insights can and often do occur between the project conceptualization, implementation, and ultimately operational phases. This is one of the learning benefits of actually doing things by reducing ideas to practice.

References

1. Energy efficiency trends in residential and commercial buildings (2008) [Online] Available at: https://www1.eere.energy.gov/buildings/publications/pdfs/corporate/bt_stateindustry.pdf (Accessed Dec 15, 2019)
2. Campbell, A. T. et al., (2008) 'The rise of people-centric sensing' *IEEE Internet Computing*, 12(4) p.12–21.
3. Froehlich, J., Findlater, L., Landay, J. (2010) 'The design of eco-feedback technology', *Proc. ACM 28th International Conference on Human Factors in Computing Systems (CHI 10)*, Atlanta, Georgia, USA.
4. F. Siero, et al. (1996) 'Changing organizational energy consumption behavior through comparative feedback' *Journal of Environmental Psychology*, vol. 16, p. 235–246.
5. Pierce, J., Paulos, E. (2012) 'Beyond energy monitors: interaction, energy, and emerging energy systems,' *Proc. ACM Conference on Human Factors in Computing Systems*, p. 665–674.
6. Foster, D. et al. (2012) 'Watts in it for me?: design implications for implementing effective energy interventions in organisations', *Proc. ACM Conference on Human Factors in Computing*, p. 2357–2366.
7. Weng, T., Agrawal, (2012) 'From buildings to smart buildings – sensing and actuation to improve energy efficiency,' *IEEE Design & Test of Computers*, 29(4), p 36–44.

8. Consolvo, S. et al (2008) 'Flowers or a robot army? Encouraging awareness & activity with personal, mobile displays', *Proceedings of the 10th International Conference on Ubiquitous Computing: UbiComp,* Seoul, S. Korea, p. 54–63.
9. Milenkovic, M. et al (2013) 'Improving user comfort and office energy efficiency with POEM (personal office energy monitor)' *Proceeding of the ACM SIGCHI Conference CHI'13 Extended Abstracts on Human Factors in Computing Systems,* p 1455–1460, Paris, France.
10. Energy Star [Online] Available at: https://www.energystar.gov/products/office_equipment/computers (Accessed Dec 15, 2019)
11. ASHRAE Standard 55-2004 – Thermal environmental conditions for human occupancy (2004) [Online] Available at: http://www.ditar.cl/archivos/Normas_ASHRAE/T0080ASHRAE-55-2004-ThermalEnviromCondiHO.pdf. (Accessed Dec 15, 2019)
12. Standard ECMA-383 'Measuring the energy consumption of personal computing products' (2010) [Online] Available at: https://www.ecma-international.org/publications/standards/Ecma-383.htm (Accessed Dec 15, 2019)
13. Dawson-Haggerty, S., et al. (2010) 'Smap: a simple measurement and actuation profile for physical information', *Proc ACM 8th conference on Embedded Networked Sensor Systems, SenSys '10,* New York, NY, USA, 210, pp 197–210.
14. MySQL [Online] Available at: https://www.mysql.com/ (Accessed Dec 15, 2019)
15. MongoDB [Online] Available at: https://www.mongodb.com/ (Accessed Dec 15, 2019)
16. Milenkovic, M. et al, (2013) 'Platform-integrated sensors and personalized sensing in smart buildings' *Proceeding of the 2nd International Conference on Sensor Networks (Sensornets),* Barcelona, Spain, p47–52.
17. A. Krioukov et al, (2011) 'A living laboratory study in personalized automated lighting control,' *ACM Buildsys 11,* Seattle, WA, USA.

Index

© Springer Nature Switzerland AG 2020
M. Milenkovic, *Internet of Things: Concepts and System Design*,
https://doi.org/10.1007/978-3-030-41346-0

Printed in the United States
by Baker & Taylor Publisher Services